Risk and Resilience Management of Water and Wastewater Systems:
Operational Guide to AWWA Standard J100

Risk and Resilience Management of Water and Wastewater Systems: Operational Guide to AWWA Standard J100

Disclaimer

Senior Editorial Manager – Book Products: Melissa Valentine
Manager – Publishing Operations: Gillian Wink
Specialist – Copyright, Trademarks, and Permissions: Peggy Tyler
Technical Editor: Dianne Beirne
Production: Innodata

Library of Congress Cataloging-in-Publication Data

Names: American Water Works Association, editor.
Title: Risk and resilience management of water and wastewater systems :
  operational guide to AWWA standard J100 / by American Water Works Association.
Other titles: Operational guide to AWWA standard J100
Description: Denver, CO : American Water Works Association, [2023] |
  Includes index. | Summary: "This operational guide to ANSI/AWWA Standard
  J100-21 Risk and Resilience Management of Water and Wastewater Systems
  incorporates new threat guidance and updated versions of the
  nonmandatory appendices into a guidance document that aligns with the
  updated standard. This guide is aimed at giving background and purpose
  to the process of completing a J100 analysis, complete with a case study
  as an example. The threats covered include contamination, cyber, flood,
  hurricane, ice storms, earthquakes, tornadoes, and wildfires"-- Provided
  by publisher.
Identifiers: LCCN 2023036747 | ISBN 9781647171216
Subjects: LCSH: Water treatment plants--Risk management--Handbooks,
  manuals, etc. | Sewage disposal plants--Management--Handbooks, manuals, etc. |
  Sewage--Purification--Safety measures--Handbooks, manuals, etc. |
  Task analysis--Standards.
Classification: LCC TD434 .R57 2023 | DDC 628.1/62068--dc23/eng/20231018
LC record available at https://lccn.loc.gov/2023036747

Printed in the United States of America

ISBN 978-1-64717-121-6

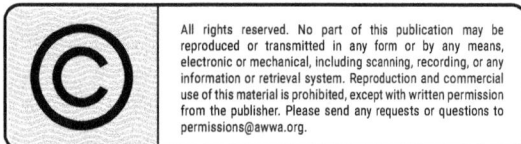

American Water Works Association

American Water Works Association
6666 West Quincy Avenue
Denver, CO 80235-3098
awwa.org

Contents

All AWWA standards follow the general format indicated subsequently. Some variations from this format may be found in a particular standard.

This page intentionally blank.

American Water Works Association

Dedicated to the World's Most Vital Resource®

Risk and Resilience Management of Water and Wastewater Systems: Operational Guide to AWWA Standard J100

SECTION 1:   ACKNOWLEDGMENTS

This Operational Guide to ANSI*/AWWA Standard J100 Risk and Resilience Management of Water and Wastewater Systems is the result of efforts to update the original ANSI/AWWA J100-10 (R13) standard. Since the initial standard was released in 2010, there has been significant discussion around the first update, especially related to the nonmandatory appendixes. To expedite the standard update process, the J100 Risk and Resilience Management Committee decided to ballot only the updated standard's mandatory sections and appendixes for the 2021 edition. The decision was made to incorporate updated versions of the nonmandatory appendixes, along with some new threat guidance, into a guidance document (operational guide) to align with the updated standard. The updated standard has been formally adopted; it is ANSI approved and referred to in this document as J100-21. This operational guide is aligned with J100-21.

---

* American National Standards Institute, 25 West 43rd Street, 4th Floor, New York, NY 10036.

The full committee led the way in this process, and its members are recognized here:

John McLaughlin, *Chair*
Jenny Graves, *Vice-Chair*

Service Provider/Consulting Services

D.B. Ballantyne, Ballantyne Consulting, LLC, Tacoma, Wash.

J.P. Brashear, The Brashear Group, Ashland, Ore.

B.O. Elwell, Elwell Consulting Group PLLC, Taylorsville, Utah

H. Gilmore, AECOM, Lewisburg, Pa.

J. Graves, AEM Corporation, Herndon, Va.

X.J. Irias (*liaison, nonvoting*), Woodard and Curran, Standards Council Liaison, Oakland, Calif.

C. Jaeger, Consultant, Albuquerque, N.M.

J.W. McLaughlin, Merrick & Company, Charlotte, N.C.

D.C. Rees, Tynwdd Consulting Group LLC, Seattle, Wash.

Management Interest

L.W. Casson, University of Pittsburgh Engineering, Pittsburgh, Pa.

S.W. Clark, USEPA, Washington, D.C.

K. Morley, AWWA, Washington, D.C.

E.S. Ralph (*liaison, nonvoting*), Standards Engineer Liaison, AWWA, Denver, Colo.

Utility/User

G.V. Burchfield, Columbus Water Works, Columbus, Ga.

N.W. Santillo, Jr., New Jersey American Water, Cherry Hill, N.J.

S.D. Spence, Prince William County Service Authority, Norwalk, Conn.

S. Stephens, Witt O'Briens, LLC., Kyle, Tex.

M. Stuhr, Portland Water Bureau, Portland, Ore.

C.T. Walczyk, SUEZ, Paramus, N.J.

For the Operational Guide specifically, the following subgroup was formed and includes standard committee members as well as noncommittee subject matter experts:

N. Anderson, Carollo, Orlando, Fla.

D.B. Ballantyne, Ballantyne Consulting, LLC, Tacoma, Wash.

J. Brados, DirectDefense, Phoenix, Ariz.

J.P. Brashear, The Brashear Group, Ashland, Ore.

A. Capelouto, Arcadis, Atlanta, Ga.

B.O. Elwell, Elwell Consulting Group PLLC, Taylorsville, Utah

J. Graves, AEM Corporation, Herndon, Va.

H. Gilmore, AECOM, Lewisburg, Pa.

A. Hurley, Critical Preparedness, LLC, Georgetown, Tex.

R. Hoye, Black & Veatch, Kansas City, Mo.

X.J. Irias, Woodard and Curran, Standards Council Liaison, Oakland, Calif.

J.W. McLaughlin, Merrick & Company, Charlotte, N.C.

K. Morley, AWWA, Washington, D.C.

P. Norton, Intera, Atlanta, Ga.

A. Ohrt, West Yost Associates, Duluth, Minn.

K. Owens, Control Cyber Inc., Spokane, Wash.

D.C. Rees, Tynwdd Consulting Group LLC, Seattle, Wash.

N.W. Santillo, Jr., New Jersey American Water, Cherry Hill, N.J.

D. Scrutchfield, Arcadis, Virginia Beach, Va.

M. Stuhr, Portland Water Bureau, Portland, Ore.

C.T. Walczyk, SUEZ, Paramus, N.J.

E. Wise, iParametrics, McDonough, Ga.

SECTION 2:   FOREWORD

AWWA's primary goal is to support water and wastewater utilities in the evaluation and improvement of their water quality, operations, maintenance, and infrastructure. Several programs and types of publications are used to support this mission.

A key program is the AWWA Standards Program, which has existed for more than 100 years to produce peer-reviewed standards for the materials and processes used by the water and wastewater utility industries. These standards, which are American National Standards Institute (ANSI) approved, are recognized worldwide and have been adopted by many utilities and organizations. The AWWA Standards Program is designed to assist water and wastewater utilities and their service providers in meeting the expectations of their customers, investors, and government regulators. The standards developed under the program are generally intended to improve a utility's overall operations and service.

Recently, AWWA also developed a series of management standards for water and wastewater utilities. The Utility Management Standards Program provides a means to assess service quality and management efficiency based on recognized standards for best available practices. Through these standards and formal recognition by professional organizations, the program serves water and wastewater utilities by promoting improvements in the quality of services and efficient management.

The utility management standards (also known as the G and J series standards) address the utility manager's need to have consistency and reliability and to know what is expected in the management and operation of a utility. These standards are also valuable resources for the many issues that utilities face, including increased scrutiny of accountability, increased regulation, and difficult economic realities such as aging infrastructure, changing water demands, and a shrinking workforce.

The utility management standards are designed to cover the principal activities of a typical water and/or wastewater utility and include the following:

- ANSI/AWWA G100, Water Treatment Plant Operation and Management,
- ANSI/AWWA G200, Distribution Systems Operation and Management,
- ANSI/AWWA G300, Source Water Protection,
- ANSI/AWWA G400, Utility Management System,
- ANSI/AWWA G410, Business Practices for Operation and Management,
- ANSI/AWWA G420, Communications and Customer Relations,
- ANSI/AWWA G430, Security Practices for Operation and Management,

- ANSI/AWWA G440, Emergency Preparedness Practices,
- ANSI/AWWA G480, Water Conservation and Efficiency Program Operation and Management,
- ANSI/AWWA G481, Reclaimed Water Program Operation and Management,
- ANSI/AWWA G485, Potable Reuse Program Operation and Management,
- ANSI/AWWA G510, Wastewater Treatment Plant Operation and Management,
- ANSI/AWWA G520, Wastewater Collection System Operation and Management, and
- ANSI/AWWA J100, Risk and Resilience Management of Water and Wastewater Systems.

The utility management standards are developed using the same formal, ANSI-recognized, AWWA-managed process. Volunteers on the standards committees establish standard practices in a uniform and appropriate format. Formal standards committees are formed to address the individual standards of practice for the diverse areas of water and wastewater utility operations.

ANSI/AWWA Standard J100, Risk and Resilience Management of Water and Wastewater Systems, is the definitive standard for water and wastewater utilities to assess and manage their risk and resilience for natural events, malevolent acts, dependency, and proximity hazards. The first edition of AWWA J100 became effective on July 1, 2010, and was revised in 2021 with an effective date of May 1, 2021. The standard identifies the seven steps required for determining a utility's existing level of risk and resilience and then managing the same.

SECTION 3:   INTRODUCTION

This operational guide is centered around Section 4, Process and Purpose, which discusses approaches for completing a J100 analysis. In other AWWA operational guides, this section is titled Requirements; however, with J100-21 the approaches presented are not the only way to perform and complete an analysis. This guide is aimed at giving background and purpose to the process of completing an analysis, complete with a case study in an appendix as an example.

With Section 4 as the explanation of the overall process and purpose of completing a J100-21 analysis through all seven steps, the appendixes that follow support that analysis. Specifically, Appendixes C through I cover the current natural hazard threats identified as part of the J100-21 reference threat library. As explained in Section 4, each of these threats needs to be considered and included, if relevant. Additionally, this guide includes Appendixes A and B covering threats that were not fully developed before the release of the original J100, including contamination and cyber.

There is a case study in Section 5. This is one approach on how to complete an analysis, using a fictitious utility named the Red River Water Department (RRWD). While this is one variation on how to complete an analysis, this and all other variations on the process must still follow the requirements in Section 4 of J100-21.

In October 2018, the federal government enacted the America's Water Infrastructure Act of 2018 (AWIA). This act contains a section that requires all water utilities serving more than 3,300 people to complete a risk and resilience analysis, using a staggered schedule based on the number of customers the utility serves. While AWIA does not require the use of the J100 standard, many utilities found J100 to be the most appropriate process to follow to assure compliance.

AWIA also requires updates to the same utility's emergency response plans to align with the findings of the risk and resilience analysis. An additional requirement included in AWIA and matching the requirements of J100-21 is to update the assessment no less than every five years. Finally, AWIA states that the (US Environmental Protection Agency) Administrator shall "recognize technical standards that are developed or adopted by third-party organizations or voluntary consensus standards bodies that carry out the objectives or activities required by this section as a means of satisfying the requirements" such as the J100 standard.

SECTION 4:   PROCESS AND PURPOSE

This section discusses the seven main steps for performing a risk and resilience analysis with the J100-21 standard (Figure 4-1). For each step in the process, this section describes the:

- **Rationale**: purpose of the step.
- **Approach**: description of potential approaches that could be used to address J100 requirements.
- **Considerations**: in developing an analysis consistent with J100.

Note that this is supplemental and background information and should not be considered a replacement of the standard or its requirements.

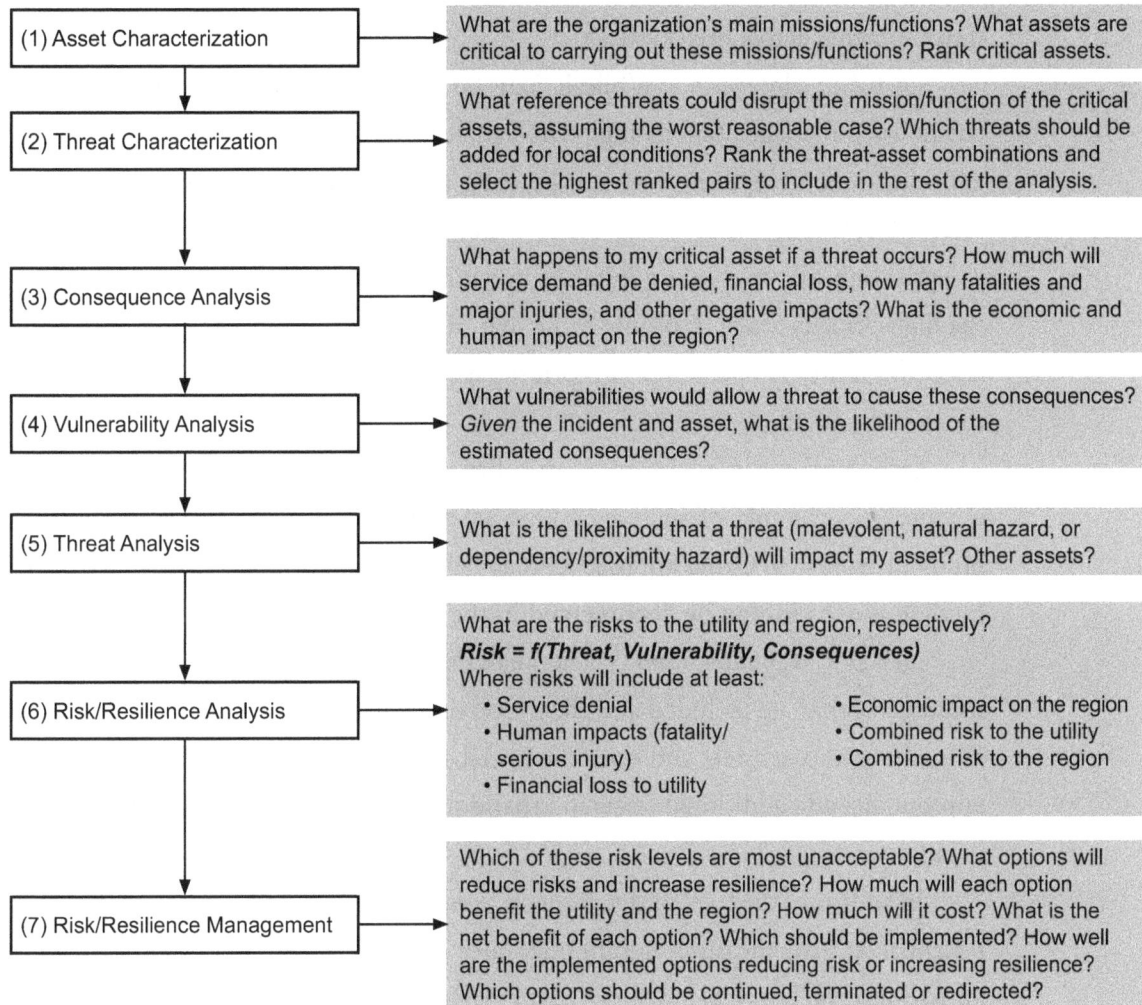

(1) Asset Characterization	What are the organization's main missions/functions? What assets are critical to carrying out these missions/functions? Rank critical assets.
(2) Threat Characterization	What reference threats could disrupt the mission/function of the critical assets, assuming the worst reasonable case? Which threats should be added for local conditions? Rank the threat-asset combinations and select the highest ranked pairs to include in the rest of the analysis.
(3) Consequence Analysis	What happens to my critical asset if a threat occurs? How much will service demand be denied, financial loss, how many fatalities and major injuries, and other negative impacts? What is the economic and human impact on the region?
(4) Vulnerability Analysis	What vulnerabilities would allow a threat to cause these consequences? *Given* the incident and asset, what is the likelihood of the estimated consequences?
(5) Threat Analysis	What is the likelihood that a threat (malevolent, natural hazard, or dependency/proximity hazard) will impact my asset? Other assets?
(6) Risk/Resilience Analysis	What are the risks to the utility and region, respectively? **Risk = f(Threat, Vulnerability, Consequences)** Where risks will include at least: • Service denial • Human impacts (fatality/ serious injury) • Financial loss to utility • Economic impact on the region • Combined risk to the utility • Combined risk to the region
(7) Risk/Resilience Management	Which of these risk levels are most unacceptable? What options will reduce risks and increase resilience? How much will each option benefit the utility and the region? How much will it cost? What is the net benefit of each option? Which should be implemented? How well are the implemented options reducing risk or increasing resilience? Which options should be continued, terminated or redirected?

Figure 4-1   The seven-step J100 process

Sec. 4.1   Asset Characterization

The first step in any risk and resilience analysis is the selection of the assets that will be included in the analysis.

4.1.1   *Rationale.* Identification of assets to be evaluated is the fundamental building block of an analysis. The utility may wish to consider assets in categories such as physical, IT, employees, customers, and knowledge base, but any grouping is acceptable. The assets that are selected for inclusion will be "threatened" by the malicious acts or natural threats defined in later steps. While only water utility assets are required for compliance with the America's Water Infrastructure Security Act of 2018 (AWIA), utilities may choose to include their wastewater and stormwater assets as well to assist in prioritization of potential improvements across their utilities.

4.1.2   *Approach.*   While it may be attractive to simply use the utility's entire asset management listing, that can lead to a large number of assets for consideration. (The term *asset* can be used to identify components of a utility's system.) Since the number of assets owned by a utility can be substantial, the analysis team may wish to undertake some initial ranking and screening to identify the high-priority assets (typically those that, if successfully attacked, would severely affect the ability to operate). High-priority assets are typically addressed first and in the greatest detail.

One approach is to assign a criticality of the assets in a qualitative ranking scheme. An example of such a ranking scheme is shown in step 1 of the Case Study in Section 5. Other approaches may be used for asset selection, but whatever method is used should be documented in the analysis.

4.1.3   *Considerations.*   The analysis gets increasingly complex with the more assets that are included and their level of detail. It may be useful to start with a smaller group of assets and once the initial analysis is complete, to review and, if appropriate, add additional assets to expand the scope of the analysis and add more detail.

In addition, the level of detail for an asset should be consistent with where the potential security measures/countermeasures may be applied. For example, having an asset level of "water plant" is not particularly useful, as potential countermeasures that may be applied would be to key parts of the plant, such as perimeter, tanks, and power supplies.

Sec. 4.2   Threat Characterization

In this second step, the threat scenarios to be used are identified and described in enough detail to estimate vulnerability and consequences.

4.2.1   *Rationale.*   Threats may be potential malevolent attacks, defined natural hazards, or dependency/proximity hazards that may adversely affect a facility or system. A description of these threat types is as follows:

- **Malevolent threats.**   Specific malevolent threat scenarios suggested by the US Department of Homeland Security (DHS), based on its characterizations of the collective activities of law enforcement, DHS, and other intelligence agencies that have developed an understanding of the means, methods, motivations, and capacities of adversarial threats to include various modes of attack with explosives (e.g., air, land, and water), various sizes of attacks (e.g., small, medium, large, and extra-large), and attacks not involving explosives (e.g., contamination, theft, and cyberattacks).

- **Natural hazards and contamination.** Hurricanes, tornadoes, floods, ice storms, earthquakes, and other events have all challenged the water sector's ability to withstand and continue to provide service. While a successful malevolent attack may be difficult for an oversight body to comprehend and understand, virtually everyone can relate to the destruction and impact of naturally occurring events. In a similar manner, the water sector, together with other sectors (food, pharmaceuticals, etc.), needed a way to address and characterize intentional or accidental contamination of their products.

- **Interdependencies and proximity hazards.** Risks due to supply chain breakdowns and collateral damage from incidents at outside sites in proximity to the utility's assets may be important to consider. For example, because of the 2001 attack on the World Trade Center, the damage to the buildings, a primary target, also severely damaged other buildings and the systems providing transportation, power, water and sanitation, telecommunications, banking, etc. Other dependency hazards, the product of cascading failures across infrastructures, require a more regional approach because the individual owner cannot be expected to be knowledgeable about these remote linkages. Proximity hazards are "dependency" threats that result from being located near hazardous sites.

4.2.2 *Approach.* Threats that should be considered are included in Table 1 of J100-21. An expanded version of this table containing additional specifications can be found in Appendix A of J100. While in some cases, the severity of a specific type of threat attack is expected to increase from left to right on the table (e.g., marine, aircraft, land-based vehicles, and assault), no such severity continuum is implied in others (e.g., theft, natural hazards) or their relative location of the threat in the table. The natural hazard threats are derived from data compiled over many years by federal agencies and are based on the physical location of the review facility. The addition of dependency and location hazards addresses the issue of being critically dependent on elements of the supply chain, especially basic infrastructures, and being located close to other assets that may pose the risk of incurring collateral damage. The utility decides which of the defined scenarios

Table 4-1   Asset/threat consequence

Assets	Consequences of Threats on Assets			
	Contamination of Water Distribution System, C {C}	Hackers Gain Control of SCADA S(CU)	100-Year Flood F1	Attack on Equipment/ Facilities AT1
Water				
Distribution Piping	2	2	3	2
Distribution Tank	9	2	3	8
Drinking Water SCADA	2	8	3	2
Chlorination	2	2	8	9
Pump Station	1	2	8	2
Wastewater				
Collection	1	2	2	2
Wastewater SCADA	2	8	2	2
Lift Stations	8	2	8	2
Prioritization Scale Used for Example	**Consequence Levels Expected**		**Numeric Range**	
	Low		1–3	
	Medium		4–6	
	High		7–8	
	Very High		9–10	

represent real, physically possible threats for the facility being evaluated; some, such as a major marine attack in a desert, may be impossible.

For those threats that are possible, the utility should assess the qualitative/ relative level of consequences of a successful attack by each threat against each asset under consideration. One approach to do this is to array a matrix of the assets versus the threats and grossly estimate consequences qualitatively according to a three- or five-point scale (e.g., very low, low, moderate, high, and very high). The utility can then examine the threat–asset pairs that are highest ranking first and proceed to lower-priority threat–asset pairs until the consequences are acceptable or the time available for the analysis is exhausted. An example of this is used in the Case Study, in Sec. 5.2.2.2, and the table is reproduced here.

4.2.3 *Considerations.* Threat characterization involves more than assuming the specific threat is applied to a specific target or asset. It requires that the analysis team considers the consequence potential for each threat and its potential to cause the maximum credible consequences, i.e., the worst reasonable case. If a threat can result in an asset causing greater consequences beyond the destruction of the asset or facility, then this combined scenario, or weaponizing of an asset, should be considered. For example, the destruction of a dam could release water downstream and inundate property below the dam. If this event were to occur at a time when the inundated area would be highly populated (e.g., on a holiday weekend), the water becomes a weapon to cause additional consequences and terror. Threat characterization requires that the assessors attempt to maximize the consequences while expending the minimum resources of the adversary. Coincidences of conditions not under the adversary's control, e.g., wind direction, should not be assumed.

Sec. 4.3   Consequence Analysis

Consequence analysis is the identification and estimation of the worst reasonable consequences generated by each specific threat–asset combination.

4.3.1   *Rationale.*   This third step examines facility design, layout, and operation to identify the types of consequences that might result. Consequences that are quantified include fatalities, service outages, serious injuries, and economic impacts.

- Fatalities and serious injuries should be displayed in presenting risks, resilience levels, and benefits in terms of the number of lives lost and injuries incurred.

- Service outage—the estimated daily outage quantity of unmet demand and duration in days.

- Regional economic impacts are widely recognized as key indicators of consequences in analyzing risks from malevolent attacks, natural hazards, and dependencies. Specifically defining the meaning of "economic impacts" is necessary for a risk management methodology.

- Financial consequences to the utility include all necessary costs to repair or replace damaged buildings and equipment, abandonment and decommissioning costs, site and environmental cleanup, revenue losses (including fines and penalties for failing to meet contractual production levels) while service is reduced, direct liabilities for casualties on and off the property, environmental damages that cannot be fully mitigated, and fines for environmental damage. These costs are reduced by applicable insurance or restoration grants and must be corrected to account for tax effects for taxpaying utilities.

- Other consequences may be identified and included in the analysis but are not required for compliance with J100-21.

4.3.2   *Approach.*   The worst reasonable consequences are estimated for each asset–threat pair for the categories being included. The standard requires that fatalities and injuries be addressed and provides a table format for consistency.

The primary concern for the public or community is the length of time, quantity, and sometimes quality of service denied and the economic consequences of service denial to the utility's direct suppliers and customers. In addition to these "direct" losses, the community suffers "indirect" losses through reduced economic activity in general, i.e., to the suppliers' suppliers and customers' customers, and so on. The economic consequences "ripple" through the regional economy, with the total impacts being some multiple of the direct impacts, hence the term *multiplier effect*. When the service denial is of short duration and/or customers can cope by such actions as conservation, substitution, redundancies, or making up lost production later through overtime or added shifts, the region is said to be "resilient." The public's objective is to enhance the resilience of critical infrastructures on which they depend.

The direct and indirect losses to the community can be calculated by a straightforward, modified input-output algorithm, originally developed to fill a gap in the computational ability of HAZUS, the Federal Emergency Management Agency's loss estimation software, and is referred to as a "HAZUS patch." The algorithm can be applied to any estimate of infrastructure service disruption to compute both the losses of output to direct facility customers and the indirect (multiplier effect) losses throughout the economy of a given region.

When a single estimate of risk, resilience, or benefits of improvements is needed for decision-making (e.g., when allocating budget resources to a large portfolio of improvements), utilities should estimate the dollar equivalent of fatalities and serious injuries. For the owner's case, the legal liabilities in excess of insurance should be used. For the metropolitan region's impact, the "value of a statistical life" should be added to the estimated regional economic impacts. If the user decides to do this, the US Environmental Protection Agency (USEPA) recommends that the central estimate of the "value of a statistical life" of $7.4 million (in 2006 dollars), updated to the base year of the analysis, be used in all benefits analyses that seek to quantify mortality risk reduction benefits regardless of the age, income, or other population characteristics of the affected population. For the current USEPA estimates, see Environmental Protection Agency, Guidelines for Preparing Economic Analyses, September 2000, EPA 240-R-00-003. The US DOT Standard as referenced in the J100-21 standard may also be used.

4.3.3   *Considerations.*   Other consequences are identified and described qualitatively, and include impact on iconic structures, governmental ability to operate, military readiness, and citizen confidence in the utility, product, or the government.

Sec. 4.4   Vulnerability Analysis

Step 4 estimates the likelihood that each specific threat or hazard, given that it occurs, will result in the worst reasonable-case consequences and will overcome the defenses of each asset included to the level identified in the consequence estimate for that threat–asset combination.

4.4.1   *Rationale.*   In the case of a malevolent attack, this means the probability that the attack would successfully result in the estimated consequences. This vulnerability is also called "probability of success" of the malevolent attack. For other hazards, it means the probability that the estimated consequences would result if the specific hazard occurred. Vulnerability analysis involves an examination of existing security capabilities and structural components, as well as countermeasures/mitigation measures and their effectiveness in reducing damages from threats and hazards.

4.4.2   *Approach.*   There are several methods that may be used to assist in estimating vulnerability. These include the following:

- **Expert judgment/direct expert elicitation.** Members of the evaluation team who are familiar with a facility's layout and workflows and are knowledgeable about the asset discuss the likelihood of success and their reasoning for their estimates. Sometimes trained facilitators, on staff or under contract, are used to elicit the judgments. In its more elaborate form, a statistical "Delphi" or Analytical Hierarchy Process can be used to establish a consensus.
- **Path analysis.** The analysis of the physical paths that adversaries can follow to accomplish their objective. The countermeasures/mitigation measures are defined by detection, delay, and response. The time required for the adversary to complete the task is compared with the estimate of the reaction time for the response force.

- **Vulnerability logic diagrams (VLDs).** The flow of events from the time an adversary approaches the facility to the terminal event in which the attack is foiled or succeeds, considering obstacles and countermeasures/ mitigation measures that must be surmounted, with each terminal event associated with a specific likelihood estimate. This is frequently complemented by time estimates for each segment and compared with an estimate of the reaction time of a counterforce once the attack has been detected. VLDs are often prepared in advance for use as heuristics to guide teams in making analyses in large or numerous facilities to enhance comparability.

- **Event trees.** The sequence of events between the initiation of the attack and the terminal event is described as a branching tree, where each "branch" represents the possible outcomes at that junction, e.g., a locked door may be breached or not. The evaluation team estimates the probability of each outcome. Multiplying the probabilities along each branch, from the initiating event to each terminal event, calculates the probability of each unique branch, while all branches together sum to unity (1.0). The sum of the probabilities of all branches on which the attack succeeds is the vulnerability estimate.

- **Fault trees.** A deductive analysis process for representing the logic combination of various system states and possible causes of that contribute to a specific event (called the top event). This approach is not commonly used for an entire analysis but may be useful in some situations and for evaluation of complex failure scenarios.

- **Hybrids of these.** Often used by the more sophisticated analysis teams.

4.4.3   *Considerations.*   Expert judgment/direct elicitation often seems to be easier and less time-consuming, but the time to reason through each threat–asset pair may lead to long discussions, and it is difficult to maintain logical consistency across several such judgments. The respective VLDs include breaking down of the overall vulnerability question into a series of implicit smaller elements that may be easier to make judgments about. VLDs have the virtue of being predefined and able to guide discussions and estimates along relevant paths efficiently and consistently. The same can be said for event or fault trees, with the added advantage that a true conditional probability is estimated, and the evaluation team is exposed to the uncertainties in their estimates.

The more structured methods (or the hybrids) may produce a more reliable estimate in the sense that a different evaluation team (or the same team at another time) is more likely to make the same or very similar estimates, given the same threat–asset scenarios and the reasoning are documented in detail. This greatly increases the consistency and direct comparability of the analyses and permits them to be used over time to measure progress of security programs or assess evolving conditions. The vulnerability estimate may be either a single-point estimate or a distribution.

Sec. 4.5   Threat Analysis

Step 5 estimates the likelihood of malevolent attack, dependency/proximity hazard, or natural hazard.

4.5.1   *Rationale.*   The threat analysis produces the probability (expressed as a positive value between 0.0 and 1.0) that a particular threat—malevolent attack, dependency hazard, or natural hazard—will occur in a given timeframe (usually one year).

4.5.2   *Approach.*   The approach differs for malevolent threats and natural threats.

- **Malevolent threats.** In general, the threat likelihood for natural threats is based on historical data, while the likelihood of malevolent threats is based on an estimate. The J100-21 standard allows two methods for malevolent threat estimates: proxy method or a "best estimate."
    - **Proxy method.** A proxy method that is well defined in the J100-21 standard.
    - **Best estimate.** It is also permissible to use the estimates made available by qualified experts ("best estimates" method). Using this method, likelihood is determined based on informed experience of the organization, input from federal, state, and local law enforcement, and others. The likelihood will be a probability with a value between 0.0 and 1.0. USEPA has developed a resource titled "Baseline Information on Malevolent Acts for Community Water Systems" that may be used to inform the best estimate approach. The utility shall document and record all assumptions, data sources, and reasoning used in estimating terrorism threat likelihood.
- **Natural hazards.** Estimates of the probability of natural hazards draw on the historical record for the specific location of the asset. Federal agencies collect and publish records for hurricanes, earthquakes, tornadoes, and

floods, which can be used as frequencies for various levels of severity of natural hazards. If there is reason to believe that the future frequency of natural hazards will differ from the past, the historical frequencies can be used as the basis for adjustments. Appendixes D through I provide guidance on estimating the frequencies of natural hazards.

- **Dependency hazards.** Initial estimates of the likelihood of dependency hazards are based upon local historical records for the frequency, severity, and duration of service denials. These estimates may serve as a baseline estimate of "business as usual," and incrementally increased if the analyst believes they may be higher due to malevolent activity on the required supply chain elements. Confidential conversations with local utilities and major suppliers of critical materials may inform these estimates.

- **Proximity hazards.** Likelihood of incurring collateral damage from an attack on a nearby asset is estimated based on the local situation and using the same logic in estimating malevolent threats.

- **Cyber threats.** Likelihood of cyber threats is 1.0 in accordance with the J100-21 Standard. See Appendix B for a rationale and example.

4.5.3   *Considerations.*   Threat likelihoods directly affect the quantitative risk results. It may be useful to perform "bounding" or "sensitivity" analysis with an alternate set of threat likelihood values to observe the impact of the threat likelihood on results.

Sec. 4.6   Risk/Resilience Analysis

Risk and resilience analysis (step 6) creates the foundation for selecting strategies and tactics to counter or mitigate disabling events by establishing priorities based on the levels of risk and resilience and the extent they can be improved.

4.6.1   *Rationale.*   The risk analysis step is a systematic and comprehensive evaluation of the previously developed estimates. Risk and resilience are estimated using the equations and methods outlined in J100-21 for all threat–asset pairs.

4.6.2   *Approach.*   This is a straightforward calculation to multiply the vulnerability times the consequences times the threat likelihood for each asset–threat pair. The results are then summed to obtain an estimate of the total risk.

4.6.3   *Considerations.*   These can become a large effort with an increased number of assets and threats. There are several software tools that can be of assistance and great value, ranging from spreadsheets to custom-tailored risk assessment software.

Sec. 4.7   Risk/Resilience Management

Step 7 addresses how changes or improvements could be made to reduce the risk, improve resilience, and enhance the reliability posture of the utility. Through the intelligent and informed management of risk, the utility positions itself to improve its level of service and security to its customers and the community.

4.7.1   *Rationale.*   The assessment provides the foundation to quantify risk in a defensible and reproducible basis for supporting resource allocation decisions (time, money, people, etc.) to reduce risk and enhance resilience. The J100-21 analysis informs the risks of water and wastewater utilities to owners/ operators and leaders of the communities they serve, supporting decision-making and resource allocation.

This step investigates potential ways to reduce risk and increase resilience. It supports the decisions to select specific countermeasure and consequence-reduction options based on the determination of an acceptable level of risk and resilience at an acceptable cost. Risk and resilience management is the deliberate process of understanding risk and deciding on and implementing action (e.g., new security countermeasures, consequence mitigation features, or characteristics of the asset) to achieve an acceptable level of risk and resilience at an acceptable cost. The initial risk and resilience analysis is based on the existing conditions at the asset.

4.7.2   *Approach.*   After this baseline risk level has been established in step 6, optional approaches to reduce risk and/or increase resilience can be defined and evaluated. The value or benefit of the options is estimated by revisiting steps 2, 3, and 4 and re-estimating the (reduced) threat likelihood, vulnerability, and/or consequences to calculate a new risk and resilience with the option in place. The reduction in risk and the increase in resilience are the benefit or value of the option, which can be compared to the cost of implementing it and to the benefits of other options. The options are classified as either countermeasures directed toward reducing threat likelihood or vulnerability, or consequence-mitigating actions, intended to reduce the economic and public health consequences of an incident and hasten a return to full functionality. Taking no action is always a baseline option against which all others are compared.

The following is one approach that can be used.

1. "Decide" what risk and resilience levels are acceptable by examining the estimated results of the first six steps for each threat–asset pair. For those that are acceptable, document the decision. For those that are not acceptable, proceed to the next steps. Not all risks and resilience

levels justify actions. This step allows the utility to decide whether it can accept the existing risk and resilience, in which case it only needs to be documented, or desire to evaluate its options for reducing the risk and enhancing resilience.

2. "Define" countermeasure and mitigation/resilience options for those threat–asset pairs that are not acceptable. Develop these alternative potential countermeasures and consequence-mitigation actions as a function of specific attack scenarios and include devalue, deter, detect, delay, and response principles as well as consequence reductions and resilience enhancements by adding steps such as redundant capabilities, continuity of operations plans, accelerated recovery, etc. In developing options, examination of the earlier estimates of consequences, vulnerability, and resilience for ways to improve them is a useful way to develop options. The following questions illustrate this concept: How can consequences be reduced? How can the asset be made less vulnerable? How can the service outage be made less severe or shorter?

3. "Estimate" the investment and operating costs of each option, being sure to include regular maintenance and periodic overhaul if expected. Adjust future costs to present value. The costs should all follow the principle of forward costing only, i.e., no previous outlays ("sunk" costs) are to be included. The only exception to this is where the user is a taxable organization, when unused depreciation can affect forward tax liabilities.

4. "Assess" the options by analyzing the facility or asset under the assumption that the option has been implemented—revisiting all affected steps 3 through 6 to re-estimate the risk and resilience levels and calculating the estimated benefits of the option (the difference between the risk and resilience levels without the option and those with the option in place). The baseline for comparison is the "do-nothing" option. The benefits are the expected value of the risk reduction after the assumed implementation (baseline risk value minus option risk value).

5. "Identify" the options that have benefits that apply to multiple threat–asset pairs. For example, if a higher fence changes the vulnerability for an attack by one assailant as well as an attack by two to four, the benefits of the two asset pairs should be added together as the benefit of the combined option. Accumulate the total benefits of each option. Once the benefits of each option of the individual threat–asset pairs

are determined, the options are examined for instances in which one option (or a design variation) reduces the risks or enhances the resilience of threat–asset pairs other than the one it was originally conceived to improve. The Case Study in Section 5 contains an example of a table/matrix that shows a comparison of different countermeasure options. Then, each option is reviewed to determine whether it would also reduce risk or increase resilience for any other threat–asset pairs, wholly or in part.

6. "Calculate" the net benefits and benefit–cost ratio (and/or other criteria that are relevant in the utility's resource decision-making) to estimate the total value and risk-reduction efficiency (benefit–cost ratio) of each option. Net benefits equal gross benefits (loss avoided) minus the present value of the costs. The benefit–cost ratio equals net benefits divided by the present value of the costs. Calculate these using the results of steps 3, 4, and 5, and add them to the table. The net benefits are the total value that each option adds, while the ratio is a direct measure of the amount of risk reduction per unit of cost—an efficiency comparison. For the economic metrics, the ratio should equal or exceed unity (one) to be considered. For fatalities and serious injuries, the ratio is the reduction in the expected number of cases per dollar, with no obvious threshold level. For some purposes, it is useful to combine fatalities with the economics for an integrated metric.

7. "Review" the options considering all the dimensions—fatalities, serious injuries, financial losses to the owner, economic losses to the community, and qualitative factors—and allocate resources to the selected options. Favor the options that have the highest net benefits, benefit–cost ratios, lives saved, and injuries avoided, considering both risk and resilience. Because the metrics are not necessarily correlated, use judgment to make the needed trade-offs. Determine the resources—financial, human, and other—needed to operate the selected options. The utility is free to assign its own weights to the respective benefits.

8. "Monitor and evaluate" the performance of the selected options. Manage the operation of the selected options, evaluate their effectiveness, and make midcourse corrections for maximum effectiveness. Implementation of options is not required by this standard.

9. "Conduct" periodic additional risk analyses to monitor progress and adapt to changing conditions. Repeat the risk analysis cycle periodically or as needed given intelligence or changing circumstances, e.g., new technologies, and new facilities. This keeps the analysis up to date and encourages continuous improvement.

The analysis in step 7 consists of the recalculation of some or all steps, which will most likely result in an overall reduced risk of threat, vulnerability, and/or the consequences of an attack. Risk reduction is recognized by comparing the current risk with the risk faced, assuming the system changes and resilience-enhancement options have been implemented. The amount of risk reduction (lowered vulnerability, reduced threat/hazard probability, or diminished consequences) or resilience enhancement (reduction in the number of days of lost service at each level of quality and the corresponding losses to the community) result in and define the benefits of the chosen options for the utility and the region, respectively.

4.7.3 *Considerations.* Some considerations for risk and resilience management are the application for decision support, documentation needed, update frequency, and documentation confidentiality and protection.

Decision support. The costs of the options are determined by the necessary investment and operating outlays. The net benefits and benefit–cost analysis or other indicator of marginal value (e.g., rate of return or return on investment) can be used to rank options for resource allocation. There are several distinct benefit metrics: fatalities avoided; injuries avoided; the utility's financial benefit–cost; the community's economic benefit–cost; and improvements in the qualitative consequences. Therefore, the choices among the options are seldom decided with a single metric until available resources are exhausted, but rather, a set of difficult trade-off decisions must be made. Some utilities apply explicit preferences to establish an initial portfolio of options and then adjust the selections as needed to balance the "portfolio" or program of risk-reduction and resilience-enhancement measures.

Once these decisions are made, risk management extends to monitoring the effectiveness and taking corrective actions as needed for any options implemented. The risk management process is the essential part of continuous all-hazards security improvement, repeated periodically (e.g., annual budget process) or as necessitated by changes in threats, vulnerabilities, consequences, technologies, or the evolving development of the utility's systems. In addition to investing in these options, risk can also be managed by acquiring insurance, entering into cooperative agreements, or simply accepting the calculated risk when it compares favorably with other operational risks such as financial or investment alternatives. Ideally, the utility would consider all these risk-reduction and resilience-enhancement options collectively as a mixed portfolio of risk and resilience management.

Documentation of assessment. The risk and resilience analysis needs to be documented as it is being performed so that it can be reviewed by others. This follows standard engineering practice, and the protocols and procedures for this will be organization specific. In addition, a final report needs to be prepared for compliance with the AWIA requirements, but the report is for utility use only and is not submitted to the USEPA. There is no standard format for the report, but at a minimum, it needs to address all the AWIA requirements.

Update of risk and resilience analysis. The risk and resilience analysis should be periodically reviewed and updated based upon changes to the threat landscape, changes in plant configuration and design, and changes in operation and major assets. AWIA also requires an update and recertification of the analysis every five years.

Documentation confidentiality and protection. The developed risk and resilience analysis contains sensitive information that should be protected from unauthorized distribution. In-house counsel for utilities can provide the best direction on how the documentation needs to be protected. Consideration may be given to having it treated as proprietary, confidential, and security sensitive, consistent with local, state, and federal laws.

SECTION 5:   APPLICATION OF THE J100-21 STANDARD CASE STUDY

Sec. 5.1   Introduction

This section provides guidance on the use of the J100-21 standard. It includes a worked case study in the application of the standard.

Sec. 5.2   Case Study

This example is intended to demonstrate the application of the standard on a hypothetical water utility facility. Note that each application of the standard has unique issues and challenges, and this demonstration analysis is not a definitive analysis but rather one approach to application of the standard.

5.2.1 *Statement of analysis objective and utility background.* The Red River Water Department (RRWD) in Red River, Ala., has decided to perform a risk and resilience analysis to support its decision-making on investments and to assist in satisfying regulatory requirements, including the America's Water Infrastructure Act of 2018 (AWIA) Risk and Resilience sections.

The objective is to perform a J100-21–consistent analysis for the RRWD.* Based upon this analysis, the utility will have identified risks in its facility, will understand the important contributors to risk, and will have additional information on the risks that can be used for decision-making on capital and operational expenditures. The utility will also have an analysis that can support the AWIA requirement for a risk and resilience analysis.

The RRWD supplies drinking water from a surface water source to the city of Red River, Ala., and collects the city's wastewater for treatment. Red River has 19,883 residents as of the last census. RRWD's mission is to provide a safe and adequate supply of water for the city's essential daily needs and proper treatment of wastewater. The department's mission statement emphasizes service reliability, protection of drinking water from contamination, and provision of water for fire protection and safety.

The utility manager's concern for performing a vulnerability analysis and risk assessment grew following certain security and emergency events during the past year. Considering these events described here, the utility leadership wants to ensure its readiness to address both natural disasters and man-made threats that could impact the department's assets and operations. Additional pertinent information includes:

- Last year, parts of the state experienced severe flooding. Nearly 12,000 residents in southeast Alabama were without water service for 4 days due to a damaged pumping station. Because portions of RRWD's water and wastewater operations are in the 100-year floodplain, the department is concerned with the potential impacts of flooding in its service area. The county's annual precipitation total has been increasing over the past decade, and the utility manager is also concerned about the increasing frequency of extreme weather events.

- Three months ago, the utility received a Water Information Sharing and Analysis Center (WaterISAC) alert from the county emergency management agency that described a suspected hacking incident at a small drinking water utility in Illinois. Hackers attempted to remotely access and disrupt the utility's Supervisory Control and Data Acquisition

---

* This example was developed by the USEPA for use in its Water Security Training courses and is adapted here to demonstrate a J100-21 analysis. Red River, Ala., and the RRWD are a hypothetical location and utility and were created for the practical exercises in this training. Any resemblance to a real city is coincidental.

(SCADA) system. Although the hacker failed to disrupt utility operations, the attack was similar to two other recently reported incidents. Because the Illinois utility used the same SCADA software and systems from the same manufacturer as RRWD, management is concerned about vulnerabilities in its own IT systems.

- Two weeks ago, a group of teenagers broke into an RRWD facility and spray-painted treatment plant buildings and storage tanks. Nothing was physically damaged or stolen, and treatment processes were apparently unaffected; customers are unaware of the break-in. However, this vandalism event raised concern about the potential for intentional contamination of the finished water and of resulting public health consequences. RRWD believes that contamination could plausibly occur by a threat actor pumping a chemical into the distribution system. Detecting and responding to the contaminant would take time, during which a significant portion of the distribution system could be impacted.

- One month ago, vandals attacked the chlorine storage tank at a neighboring utility and damaged the connection header for two 150-lb chlorine cylinders, resulting in fast release of chlorine gas. As RRWD has a similar chlorine treatment arrangement but with larger cylinders (two 1-ton cylinders), RRWD management wants to better understand the potential impact of such an attack at their facility.

This analysis will be an all-hazards risk assessment, performed in accordance with the J100-21 standard and able to support the AWIA requirements for a risk and resilience analysis. Although RRWD completed a vulnerability assessment in 2005 as required by the Bioterrorism Response Act, it is now outdated. Thus, this assessment will start with a clean slate. Table 5-1 summarizes information about RRWD's operations and priorities for the assessment.

5.2.2 *Analysis steps.* Referring to the background information in Table 5-1, begin an analysis file for RRWD. The J100-21 standard steps will be followed and are listed here.

1. Asset Characterization
2. Threat Characterization
3. Consequence Analysis
4. Vulnerability Analysis
5. Threat Analysis
6. Risk/Resilience Analysis
7. Risk/Resilience Management

5.2.2.1 Step 1—asset characterization.   In this step, the purpose is to answer the question: "What assets does the utility have and which are critical?" The RRWD major assets are listed in Table 5-1. If desired, all assets that the utility has may be included in the analysis. However, that approach can result in a large asset list and a corresponding large amount of analysis as each asset will need to be analyzed. The standard provides a suggested screening approach to assist in identifying the critical assets. Table 5-2 shows the specific elements of this step.

Table 5-1   Red River Water Department (RRWD) data summary

RRWD Utility Profile	
Utility supervisor	Jim Griffin
Address	101 Main Street, Red River, AL 36401
Service area	28.9 mi^2
Number of retail customers served	19,883
Number of wholesale customers served	1,000
Number of connections, residential	7,207
Number of connections, nonresidential	592
Number of connections, wholesale	1
Distribution system (mi)	174
Average daily water service (MGD)	3
Average residential water service (MGD)	1.87
Average nonresidential water service (MGD)	1.09
Average daily wholesale water service (MGD)	0.08
Average daily wastewater service (MGD)	3
Percent of wastewater flow from residential sources (%)	63
Key Assets	
Assets of concern	• Water distribution system ("Distribution piping") • Water distribution storage tank ("Storage tank") • Drinking water pumping station ("Pump station") • Chlorination system ("Chlorination") • Wastewater lift station ("Lift station") • Drinking water SCADA system ("SCADA") • Wastewater collection system Plant Process Control & Monitoring ("Wastewater SCADA")

(continued)

Table 5-1   Red River Water Department (RRWD) data summary (*Continued*)

RRWD Utility Profile	
	Note: AWIA asset categories are pipes and constructed conveyances, physical barriers, source water, pipes and constructed conveyances, water collection and intakes, pretreatment and treatments, storage and distribution facilities, and electronic computer or other automated systems. Even though the wastewater system and its components are not required by AWIA, they are being included in this analysis at management direction.
Equipment, plans, and procedures in place that may be useful in protection of assets (i.e., countermeasures)	• Emergency response plan ("Emergency operating procedural plan") • Hardened doors on drinking water pump station and lift station ("Hardened doors") • Chlorine monitoring system ("Toxicity monitoring") • IT requires antivirus and antispyware software installation on all office computers. • IT has implemented secondary user ID and password requirements for changes made to all computer systems.

Table 5-2   Analysis step 1—asset characterization

J100-21 Step Requirements		Case Study Application
4.1.1	Define the mission and critical functions of the organization.	RRWD's mission is to provide a safe and adequate supply of water for the city's essential daily needs and proper treatment of wastewater. The department's mission statement emphasizes service reliability, protection of drinking water from contamination, and provision of water for fire protection and safety.
4.1.2	Identify the assets that perform or support the mission or critical functions.	Based on the mission statement, a critical asset list was developed and is stated below.
4.1.3	List potentially critical assets.	The list of potentially critical assets at this point is the list of all the assets being considered. • Water distribution system ("Distribution piping") • Water distribution tank ("Distribution tank") • Drinking water SCADA system • Chlorination system ("Chlorination") • Drinking water pumping station ("Pump station") • Wastewater collection • Wastewater SCADA system • Wastewaters lift stations ("Lift stations")

(*continued*)

Table 5-2   Analysis step 1—asset characterization (*Continued*)

J100-21 Step Requirements	Case Study Application
4.1.4 Identify the critical internal and external supporting infrastructures.	Electric power supply is a supporting infrastructure, but the analysis of it is considered outside this analysis.
4.1.5 Identify and document existing protective countermeasures and mitigation measures/features that protect or mitigate the risks to critical assets, infrastructure, or facilities.	The existing countermeasures are listed as follows. • Emergency response plan ("Emergency operating procedural plan") • Hardened doors on drinking water pump station and lift station ("Hardened doors") • Chlorine monitoring system ("Toxicity monitoring") • IT requires antivirus and antispyware software installation on all office computers. • IT has implemented secondary user ID and password requirements for changes made to all computer systems. • For each countermeasure, a descriptive sheet similar to that shown in Table 5-3 has been prepared.
4.1.6 Estimate the worst reasonable consequences resulting from the destruction or loss of each asset, without regard to the threat. The consequences include the potential for fatalities, serious injuries, financial loss to the utility, and economic loss to the regional economy. Other consequences may also be considered such as impacts to the environment, loss of public confidence, and/or inhibiting the effective function of national defense or civilian government at any level.	Table 5-4 includes an estimate of the worst reasonable consequences of the assets on a scale of 1 to 10, with 10 being the most severe impact, for the consequence categories listed. Alternate methods for determining the steps may be used, but whatever method is used, it must be described/documented.
4.1.7 Prioritize the critical assets using the estimated consequences from 4.1.6.	As shown in Table 5-4, the prioritization was performed using a simple ranking of 1 to 10 for each consequence category, including fatalities, injuries, utility (impact), regional (impact), environment, public confidence, and defense, for each asset. In this example, it was determined that all assets would be included in the analysis. In the use of the standard, not all categories need to be used, and user may include other categories. In addition, the user may choose to only include assets that have a "score" greater than some pre-defined value.

Table 5-3 Existing countermeasure (CM) fact sheet

Chlorine monitoring system	
CM Protection:	Detection
Type of CM:	Automated—Water
Description:	Installation of sensors to monitor key elements of the process to identify anomalies before they become serious.
Assigned Assets:	Water distribution system
Emergency operating procedural plan	
CM Protection:	Response
Type of CM:	Policies and procedures
Description:	A set of procedures that define employee responses to specific type of CMs of emergency events.
Assigned Assets:	All
Hardened doors	
CM Protection:	Delay
Type of CM:	Locks
Description:	Hardened locks on doors. Intended to protect a door from being forcefully entered. Security of the doorway was enhanced by modifying the door, the door frame, the hinges, or the lock. Different doorway security measures may protect against various potential threats, including breaking, blasting, or fire.
Assigned Assets:	Drinking water pump station; wastewater lift station
Secondary user ID and password	
CM Protection:	Detection
Type of CM:	IT
Description:	Software design to ensure at least two people are aware of changes being made to critical IT programs.
Assigned Assets:	Water system—SCADA; Wastewater system—SCADA
Antivirus and pest eradication software	
CM Protection:	Detection
Type of CM:	IT
Description:	Designed to detect electronic threats to a computer or other electronic system and to delay these threats from damaging the system. In addition, some antivirus software responds to threats by deleting them or otherwise disabling them.
Assigned Assets:	Water—SCADA; wastewater—SCADA

Table 5-4   Consequence prioritization matrix

Case Study Consequence Prioritization							
	Fatalities	Injuries	Utility	Region	Environment	Public Confidence	Defense	Overall (Highest)
Water								
Distribution Piping	8	2	8	3	5	5	2	8
Distribution Tank	1	1	8	3	5	7	1	8
Drinking Water SCADA	1	2	8	5	4	6	7	8
Chlorination	9	9	6	4	7	7	6	9
Pump Station	6	3	9	4	1	7	6	9
Wastewater								
Collection	1	4	7	6	9	8	3	9
Wastewater SCADA	2	2	7	5	6	8	1	8
Lift Stations	3	3	9	6	7	7	3	9

Worst Case/Prioritization Scale Used for Example	Worst Reasonable Consequence Level	Numeric Range				
	Low	1–3				
	Medium	4–6				
	High	7–8				
	Very High	9–10				

5.2.2.2   Step 2—threat characterization.   This step is intended to answer the question: "What threats and hazards should I consider?"

After review of the J100-21 reference threats in Table 1 of the standard, discussion with plant management, and review of the history of events that the utility has faced, the utility has decided to consider the threats as shown in Table 5-5. It is recognized that additional threats may be viable, and those may be reviewed and/or analyzed in the future.

The specific requirements of this step in the standard are shown in Table 5-6.

Table 5-5   Threats to be considered by RRWD in case study

Threats to consider in example analysis:	• Contamination of water distribution system by chemical substance ("Contamination of the Product; C(C)—Contamination-Chemical") • Hackers gaining control of or disrupting water and wastewater SCADA systems ("Cyberattack; (C1)—Cyberattack") • 100-year flood impacting drinking water pumping station or wastewater lift station ("Flood; F1—100-year flood") • Attack on chlorine treatment system—intended to cause release of chlorine gas ("Assault Teams; AT1—Assault Team 1")

Table 5-6   Analysis step 2—threat characterization

	J100-21 Step Requirements	Case Study Application
4.2.1	The utility shall conduct a threat characterization using the reference threats described previously and the following:	Threats included are those shown previously. The details of each threat are shown in Table 5-7 Detailed Threat Listing. This detail was derived from the J100-21 standard and reviewed/updated by the utility.
4.2.2	Utilities shall describe for natural hazards all earthquakes, hurricanes, tornadoes, floods, ice storms, and wildfires that have occurred or could occur in the location of the facility. Natural hazards typically range in magnitude, e.g., categories of hurricanes. Define the lowest magnitude of each hazard that poses potentially unacceptable risks to continued operations. Appendix A specifies the basic event descriptions.	The example utility elected to include one natural threat, the 100-year flood, as described in Table 5-5. Also, see Appendix D—Flood.
4.2.3	Define the range of the magnitudes from the smallest that would cause serious harm to the largest reasonable case. These threats should be included in the analysis.	These threat magnitudes are defined from the selected threats using Table 1 Malevolent Threat by Category in the J100-21 standard. The threats selected have the following designation: C(C); (C1); F1; AT1.
4.2.4	Utilities shall describe, for dependency hazards, all interruptions of utilities, suppliers, employees, customers, and transportation, and proximity to dangerous neighboring sites.	No dependency hazards were identified in the example analysis.

(continued)

Table 5-6   Analysis step 2—threat characterization (*Continued*)

J100-21 Step Requirements	Case Study Application
4.2.5 Identify the threats to be included in the analysis, which apply to each asset to determine the set of threat–asset pairs. The utility may rank the threat–asset pairs according to the judged magnitude of the resulting consequences and then select the critical threat–asset pairs to be included in the rest of the analysis process or choose to evaluate all threat–asset pairs. An example method is to array in a matrix: (a) all assets selected in Sec. 4.1.7 against (b) all threats defined in Sec. 4.2.1 through Sec. 4.2.4 and enter a qualitative judgment ("small," "medium," or "large" may suffice, or a more differentiated scale, e.g., 1 through 10, may be used) as to the rough magnitude of the consequences. In general, these threat–asset pairs are the objects of analysis throughout the rest of the process.	The utility prioritized all the identified threat–asset pairs and ranked them as shown in Table 5-8 using a scale of 1 to 10. For this analysis, the critical threat–asset pairs selected for analysis are those with a prioritization ranking of 8 or higher. The other threat–asset pairs may be evaluated in the future.
4.2.6 Record the methods, assumptions, and the list of threats to be included in the analyses that follow.	The methods, assumptions, and list of threats are documented in the preceding steps.

Table 5-7   Detailed threat listing—threats used in example analysis

AT1 – Assault Team 1	
Class	Man-made threats
Type	Assault teams
Description	One assailant; by land—pedestrian, all-terrain vehicle, motorcycle, over-the-road personnel transport, cargo truck; by water—lone swimmer
Access	Pistol, assault rifle, light machine gun, grenades (H.E. and incendiary); explosive vest and/or satchel
Category	Terrorist AT1
Objectives	Disrupt utility mission—maximize loss of mission capability, loss of life and/or economic damage. Cause fear on the national/local level
Equipment	Power tools, hand tools, minimal breaching tools, lone swimmer. Transport—foot, vehicle, marine pedestrian, all-terrain vehicle (ATV), motorcycle, over-the-road personnel transport, and cargo truck
Knowledge	Extensive knowledge of water utility operations, facility layouts, and security system. Highly skilled and trained, well financed, and very good support structure. Conducts extensive planning
Motive	Terror, destroy assets
Personnel	1
Assigned Assets	Water—Chlorination, Water—Distribution tank

(*continued*)

Table 5-7 Detailed threat listing—threats used in example analysis (*Continued*)

C(C) – Contamination—Chemical	
Class	Man-made threats
Type	Contamination of product
Description	Contamination of the water distribution system with chemical
Access	Acquisition of chemical through purchase, theft, or synthesis; pistol, assault rifle, light machine gun, and grenades (H.E. and incendiary)
Category	C(C) Chemical
Objectives	Contamination of water distribution system to cause deaths, injuries, and/or economic damage
Equipment	Hand tools, tanks, pumps, mixers, and transport—foot, vehicle, cargo truck, and tanker truck
Knowledge	Minimal knowledge of water utility operations, facility layouts, and distribution system. Minimally skilled and trained
Motive	Terror
Personnel	1–4
Assigned Assets	Water—Distribution tank, Wastewater—Lift stations

F1 – 100-year Flood	
Class	Natural threats
Type	Floods
Description	Flood events that have 0.2% chance of occurring in a given year based on historical events.
Assigned Assets	Water—Chlorination, Water—Pump station, Wastewater—Lift stations

(C1) – Cyber Attack	
Class	Man-made threats
Type	Process sabotage
Description	Unauthorized access through cyber intrusion to cause harm by damaging, disabling, or destroying process control systems
Access	Information technology equipment
Category	C1—Cyber Attack
Objectives	Sabotage by causing harm by damaging, disabling, or destroying process control systems
Equipment	Information technology equipment. Will attempt to perform act from off-site if possible
Knowledge	Extensive knowledge of water utility operations and facility layouts. Highly skilled and trained
Motive	Sabotage, unauthorized access
Personnel	1
Assigned Assets	Water—SCADA, Wastewater—SCADA

Table 5-8   Consequences of threats on assets

Assets	Consequences of Threats on Assets			
	Contamination of Water Distribution System, C (C)	Hackers Gain Control of SCADA S(CU)	100-Year Flood F1	Attack on Equipment/ Facilities AT1
Water				
Distribution Piping	2	2	3	2
Distribution Tank	9	2	3	8
Drinking Water SCADA	2	8	3	2
Chlorination	2	2	8	9
Pump Station	1	2	8	2
Wastewater				
Collection	1	2	2	2
Wastewater SCADA	2	8	2	2
Lift Stations	8	2	8	2
	Consequence Levels Expected		Numeric Range	
	Low		1–3	
Prioritization Scale Used for Example	Medium		4–6	
	High		7–8	
	Very High		9–10	

5.2.2.3   Step 3—consequence analysis.   The purpose of this step is to answer the questions: "What happens to the utility's assets if a threat or hazard happens? How much money lost, how many lives lost, and how many injuries?"

The requirements for performing the consequence step and the example analysis approach are shown in Table 5-9.

Table 5-9   Analysis step 3—consequence analysis

J100-21 Step Requirements	Case Study Application
4.3.1 Apply "worst reasonable case" assumptions for each threat scenario. For malevolent threats, assume the adversary to be intelligent and adaptive; knowledgeable about utility structure, operations, and processes; and attempting to optimize or maximize the consequences of a particular attack scenario. For natural hazards, assume all reasonable event magnitudes. For dependency hazards, assume complete loss of the factor (utilities, chemicals, customers, etc.) for long enough to disrupt the utility's functioning. For proximity threats, assume the worst reasonable event to the nearby site based on the nature of the site. Do not assume that all uncontrollable variables (e.g., wind speed and direction) and unpredictable events occur simultaneously. Prepare a copy of Table 2 for each threat–asset pair to be analyzed. Define and document the assumptions used for worst reasonable cases.	Several methods for consequence analysis are available, including but not limited to the following: • WHEAT • HAZUS MH • TEVA-SPOT • Existing analysis • Historical records • Best estimate For this example, best estimate was used. The analysis for this case study shows the filling out of the J100 Table 2 as specified in this step is shown in Table 5-10.
4.3.2 Estimate the consequences of the threat to the asset as listed in Sec. 4.2 under the worst reasonable case assumption according to the following: 4.3.2.1 Estimate serious injuries and/or illnesses as the number of persons, both employees and nonemployees, on and off-site, who are injured or sickened in the threat–asset incident, including those who expire more than 30 days after the incident, using the severity levels in Table 2. 4.3.2.2 Estimate fatalities as the estimated number of persons, both employees and nonemployees, on and off-site, who are fatally injured or sickened due to the threat–asset event and die immediately or within 30 days of the event. The estimates can be recorded in a format similar to that provided in Table 2. Add the right two columns, the number of individual casualty cases and the severity-weighted number of casualties.	The estimated fatalities and injuries are shown in Table 5-10. Note that a value of $7,800,000 was used as the monetary value for a statistical life (VSL). The statistical life value is based on USEPA "Guidelines for Preparing Economic Analysis," dated Dec. 17, 2010. This value is based on a distribution fitted to 26 published VSL estimates that USEPA reviewed. The Value of a Statistical Injury is shown as calculated in the table, based on the injuries being Abbreviated Injury Scale (AIS) Level 3, which has a weight of 1.05% of the default VSL, as shown in Table 2 of J100-21. Other values for VSL and weights for injuries may be used by the utility if they are documented.

(continued)

Table 5-9   Analysis step 3—consequence analysis (*Continued*)

J100-21 Step Requirements		Case Study Application
4.3.2.3	Estimate service outage: Inability to provide the level of service needed to sustain quality, quantity, reliability, and environmental standards to meet customer needs. The specific criteria defining service outage are set by each utility (e.g., the quantity, quality, and pressure). Service outage is estimated as the product of the daily quantity of unmet demand (at specified quality, pressure, etc.). Service-denial risk is the expected value of service denial weighted by threat likelihood and vulnerability. Reducing service denial risk, by definition, increases resilience because the resilience objective is zero interruption in service.	There was no service loss estimated due to these threats on the assets considered.
4.3.2.4	Estimate owner's financial loss including the direct cost of repairing, rehabilitating, or replacing damaged plant, pipe network, and equipment, abandonment costs (if any), and other fixed-cost losses; incremental lost revenue (less any reductions in operating costs due to the outage) and legal liabilities for damages to others (net of insurance payout); environmental remediation and liabilities; and fees or penalties for failure to fulfill contracts with customers, suppliers, employees, or others, and fines for failing to maintain regulatory compliance. For utilities subject to income taxes, using after-tax calculations will yield better estimates of owner's financial losses and option costs for use in decision-making than pre-tax calculations. If these consequences extend for more than the current 12-month period, they should be discounted to present value.	This is shown in Table 5-12 Combined Consequence Analysis for all categories.

(continued)

Table 5-9   Analysis step 3—consequence analysis (*Continued*)

J100-21 Step Requirements	Case Study Application
4.3.2.5 Estimate economic losses to the regional economy as reductions in the gross regional product (regional equivalent of the Gross National Product, a measure of total economic activity on a value-added basis). Lost gross regional product can be estimated using a regional input-output model such as the RIMS II tool developed by the US Bureau of Economic Analysis. These estimate regional losses based on service denial (times average price of service) as they impact all their customers and the change in customer's activities affect their customers and so on, capturing full economic multiplier effects. As with owner's loss, if these costs extend for more than one year, the present value should be used.	This is shown in Table 5-12 Combined Consequence Analysis for all categories.
4.3.3 Combined Consequence Losses. While the level of loss of each respective consequence is important, there are also times when consolidating to a single number representing total of all losses for the relevant decision is useful for decision-makers (e.g., in comparing overall risks, risk-reduction benefits, or benefit/cost ratios over a large number of options). To combine the respective terms requires that human casualties be converted to dollar amounts to be added to the other metrics, which are naturally in dollars. The two most widely used approaches relevant to this standard are: (1) the human capital approach and (2) the value of a statistical life approach.	The Combined Consequence losses are shown in the last column of Table 5-12. The number of potential fatalities and injuries are also shown in the table. The value of a statistical life approach was used as described previously.

Table 5-10   Worst-case consequence analysis

Asset/Threat Pair	AIS Level	Severity	Examples	Weight of Lifetime Cost or VSL	Direct Cost ($)	Estimated No. of Cases	Weighted Total Casualties	NIOSH Weighted Total Casualties ($)	VSL Weighted Total Casualties ($)
Water: Chlorination; F1—100-Year Flood	0	None							
Water: Distribution Tank; C(C)—Contamination-Chemical	0	None							
Water: Chlorination; AT1—Assault Team 1	7	Immediate fatalities	1 fatality	1.0	7,800,000	1			7,800,000
Water: Chlorination; AT1—Assault Team 1	3	Serious Injuries	15 injuries estimated	0.105	7,800,000	15			12,285,000
Water: Distribution System; AT1—Assault Team 1	0	None							
Wastewater: Lift Station; C(C)—Contamination-Chemical	0	None							
Wastewater: Lift Station; F1—100-Year Flood	3	Serious injuries	20 injuries estimated	0.105	7,800,000	20			16,380,000
Water: Pump Station; F1—100-Year Flood	0	None							
Water: SCADA; (C1)—Cyberattack	0	None							
Wastewater: SCADA; (C1)—Cyberattack	0	None							

Table 5-11   Reference J100-21 Table 2. Serious injury/illness and fatalities factors

AIS Level	Severity	Examples	Weight of Lifetime Cost or VSL	Direct Cost ($)	Estimated No. of Cases	Weighted Total Casualties	NIOSH Weighted Total Casualties ($)	VSL Weighted Total Casualties ($)
1	Minor	Superficial abrasion or laceration of skin; digit sprain; first degree burn; and head trauma with headache or dizziness (no other neurological signs)						
2	Moderate	Major abrasion or laceration of skin; cerebral concussion (unconscious less than 15 min); finger or toe crush/amputation; and closed pelvic fracture with or without dislocation	0.047	5				
3	Serious	Major nerve laceration; multiple rib fracture (but without flail chest); abdominal organ contusion; and hand, foot, or arm crush/amputation	0.105	15				

(continued)

Table 5-11   Reference J100-21 Table 2. Serious injury/illness and fatalities factors (*Continued*)

AIS Level	Severity	Examples	Weight of Lifetime Cost or VSL	Direct Cost ($)	Estimated No. of Cases	Weighted Total Casualties	NIOSH Weighted Total Casualties ($)	VSL Weighted Total Casualties ($)
4	Severe	Severe spleen rupture; leg crush; chest-wall perforation; and cerebral concussion with other neurological signs (unconscious less than 24 h)	0.266	50				
5	Critical	Critical spinal cord injury (with cord transection); extensive second- or third-degree burns; and cerebral concussion with severe neurological signs (unconscious more than 24 h)	0.593	100				
Subtotals, Injuries, and Illnesses								
6	Unsurvivable	Unsurvivable injuries which, although not fatal within the first 30 days after an accident, ultimately result in death	1	125				
7	Immediate fatalities	Death incurred immediately to within 30 days as direct result of the incident	1	100				
Subtotal, Fatalities								
Totals, Casualties								

Source: Moran, M. and Monje, C. 2016. Guidance on the Treatment of Economic Value of a Statistical Life (VSL) in US DOT Analyses.

Table 5-12   Combined consequence analysis

			Consequences						
System	Asset	Threat	Number of Fatalities	Number of Injuries	Fatalities ($)	Injuries ($)	Utility Financial Impact ($)	Regional Economic Impact ($)	Combined Consequences ($)
Water	Chlorination	F1—100-year Flood	0	0	0	0	40,000	50,000	90,000
Water	Distribution Tank	C(C)—Contamination-Chemical	0	0	0	0	50,000	0	50,000
Water	Chlorination	AT1—Assault Team 1	1	15	7,800,000	12,285,000	100,000	0	20,185,000
Water	Distribution Tank	AT1—Assault Team 1	0	0	0	0	60,000	0	60,000
Wastewater	Lift Stations	C(C)—Contamination-Chemical	0	0	0	0	200,000	0	200,000
Wastewater	Lift Stations	F1—100-year Flood	0	20	0	16,380,000	50,000	100,000	16,530,000
Water	Pump Station	F1—100-year Flood	0	0	0	0	40,000	50,000	90,000
Water	SCADA	C1—Cyberattack	0	0	0	0	80,000	100,000	180,000
Wastewater	SCADA	C1—Cyberattack	0	0	0	0	60,000	100,000	160,000

5.2.2.4 Step 4—vulnerability analysis. The purpose of the vulnerability analysis is to answer the question: "What are my vulnerabilities that would allow a threat or hazard to cause these consequences?"

Table 5-13   Analysis step 4—vulnerability analysis

J100-21 Step Requirements	Case Study Application
4.4.1 Review pertinent details of the facility construction, systems, and layout. Include all existing countermeasures, mitigation measures, and other impediments to threats, such as topographic, design, and equipment features that provide deterrence, detection systems, and delay features, and local and supporting response measures. Include information on interdependencies, personnel interactions, and process flows within the facility. Identify vulnerabilities or weaknesses in the protection system.	The analysis staff reviewed the documentation and held meetings with appropriate planning and engineering staff and management.
4.4.2 Analyze the vulnerability of each critical asset or system to estimate the likelihood that, given the occurrence of a threat, the consequences estimated in Sec. 4.3 result. The utility may use fault-tree analysis, event-tree analysis, path analysis, vulnerability logic diagrams, computer simulation methods, expert judgments, or other generally recognized approaches for vulnerability determination.	Results of the vulnerability analysis for each asset–threat combination are shown in Vulnerability Table 5-14. This example vulnerability determination was performed using a qualitative evaluation of the ability to detect, delay, and respond to the analysis. In this approach, users evaluate the assigned countermeasures for each asset/threat pair and then assess the aggregate capabilities of those countermeasures for each aspect of consequence reduction: Detection, Delay, and Response, with a resulting value between 0 and 1.0. Other methods could be used, including those listed to the left.
4.4.3 Set and evaluate cyber vulnerability levels. Most utilities have automated systems that support their operations control and their business enterprise needs, respectively, both of which should have protective controls. Utilities shall assess which controls are most applicable to business enterprise systems (e.g., billing and finance, communications, and personnel), and operations control systems (e.g., monitoring, controlling processes, and automatic emergency response). At a minimum, this should consider controls defined in the NIST Cybersecurity Framework v1.1 (www.nist.gov), which can be assessed directly or by using the AWWA Cybersecurity Assessment Tool (www. awwa.org). The AWWA Water Sector Cybersecurity	The cyber vulnerability was evaluated using the AWWA Water Sector Cybersecurity Risk Management Guidelines.

(continued)

Table 5-13   Analysis step 4—vulnerability analysis (*Continued*)

J100-21 Step Requirements		Case Study Application
	Risk Management Guidance allows a utility to prepare a self-assessment of the "as-built" status of critical cyber security controls and determine a level of maturity. Based on the level of implementation for priority controls, the utility can decide on degree of vulnerability.	
4.4.4	Document the specific assumptions and procedures used for performing this consequence analysis, the worst-reasonable-case assumptions, and the results of the consequence analysis.	The analysis results for vulnerability are shown in Table 5-14.
4.4.5	Record the vulnerability values for use in Sec. 4.6 using point estimate, ranges, or probability distributions.	See Table 5-14. For convenience, the consequence categories are provided for reference.

For this example, the vulnerability, or likelihood of consequences or damages, is a function of the combined effectiveness of each countermeasure. This efficacy is determined by independently evaluating the capabilities of these countermeasures with respect to each asset–threat pair.

There are four possible levels for each capability. These evaluations are combined into a single score, whereby likelihood of consequences or damages is calculated based on numerical values assigned to each of the evaluation countermeasure capability levels. The vulnerability for man-made threats is determined by calculating the *product* of the numerical value in each of the three capability areas: detection, delay, and response (Figure 5-1a). The user may apply the combined value calculated or enter their estimate as a percentage.

The vulnerability for dependency/proximity and natural disaster threats is determined by calculating the *sum* of the numerical value in each of the three capability areas: preparation, active response, and recovery (Figure 5-1b). The user may apply the combined value calculated or enter their estimate as a percentage.

Combined countermeasure capability	
Low (0)	100%
Low (1)	90%
Low (2)	80%
Moderate (4)	75%
Moderate (6)	60%
Moderate (8)	50%
Moderate (12)	40%
High (16)	30%
High (24)	20%
High (32)	12%
Very High (48)	9%
Very High (64)	6%
Very High (96)	3%

$$\Big] = \Pi \Big[$$

Countermeasure capabilities Man-made threats		
Detection	Delay	Response
None (0)	None (1)	None (0)
Possible(1)	Limited (2)	Slow (1)
Probable (2)	Strong (4)	Variable (2)
Certain (4)	Very Strong (6)	Fast (4)

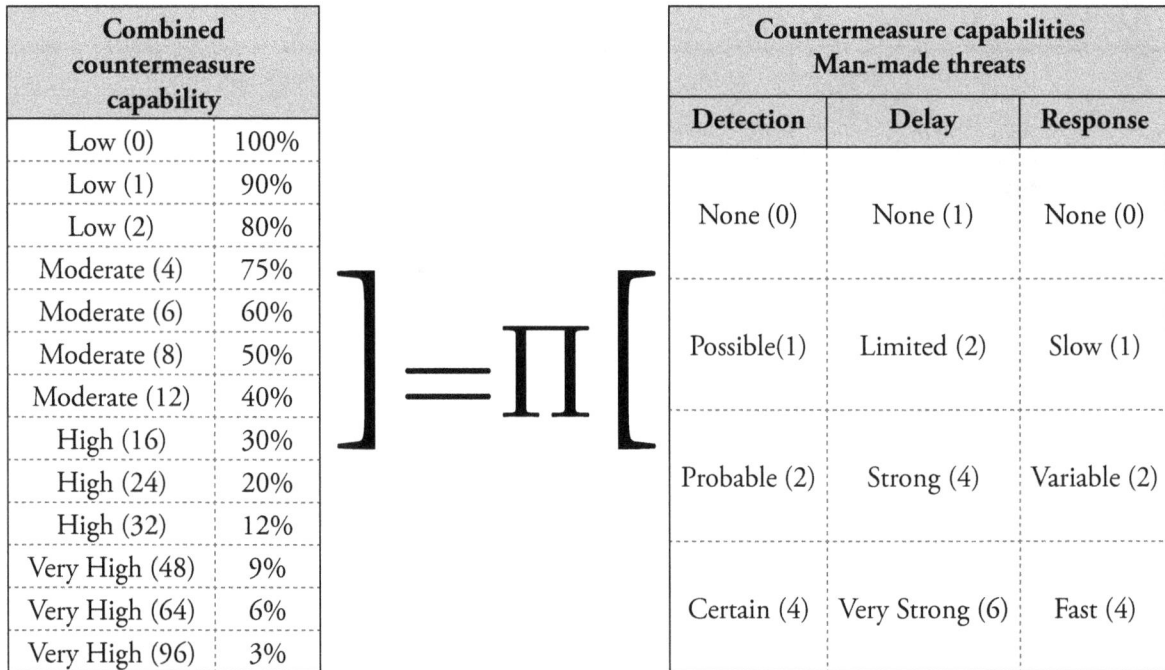

Source: Created using the USEPA online Vulnerability Self-Assessment Tool (VSAT)

5-1a   Determination of combined countermeasure capability for man-made threats

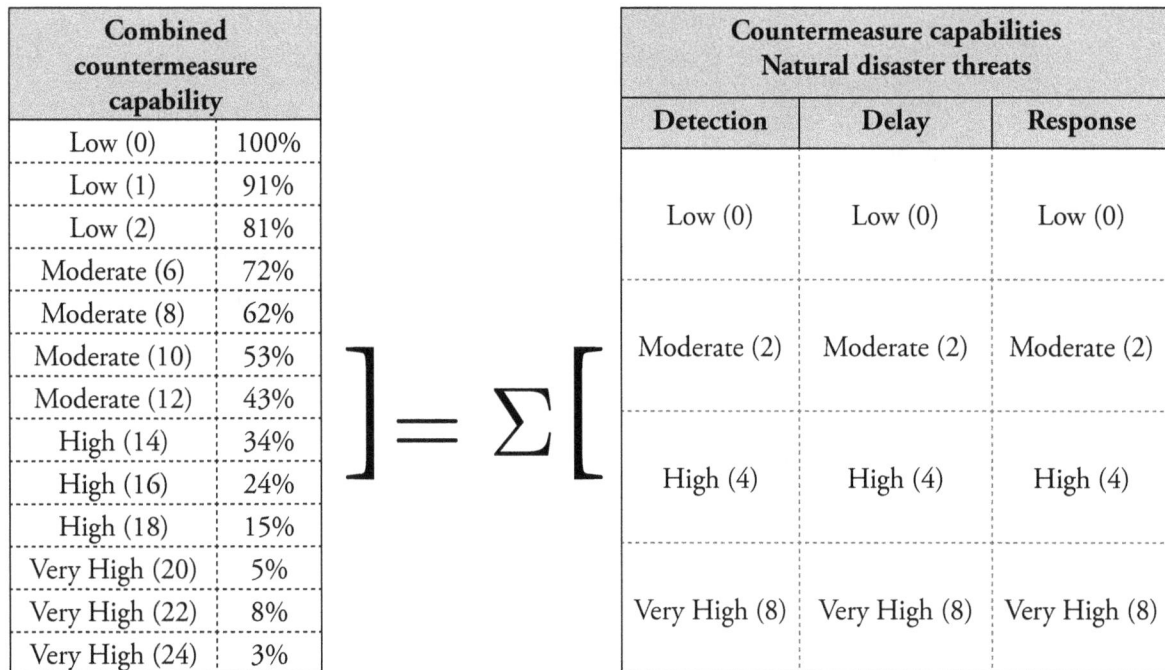

Combined countermeasure capability	
Low (0)	100%
Low (1)	91%
Low (2)	81%
Moderate (6)	72%
Moderate (8)	62%
Moderate (10)	53%
Moderate (12)	43%
High (14)	34%
High (16)	24%
High (18)	15%
Very High (20)	5%
Very High (22)	8%
Very High (24)	3%

$$\Big] = \Sigma \Big[$$

Countermeasure capabilities Natural disaster threats		
Detection	Delay	Response
Low (0)	Low (0)	Low (0)
Moderate (2)	Moderate (2)	Moderate (2)
High (4)	High (4)	High (4)
Very High (8)	Very High (8)	Very High (8)

Source: Created using the USEPA online Vulnerability Self-Assessment Tool (VSAT)

5-1b   Determination of combined countermeasure capability for dependency, proximity, and natural threats

Figure 5-1   Vulnerability calculation

Table 5-14 Vulnerability

System	Asset	Threat	Detection/ Preparation	Delay/Active Response	Response/ Recovery	Likelihood of Damage
			Vulnerability			
Water	Chlorination	F1—100-year flood	Moderate	Moderate	Moderate	0.72
Water	Distribution Tank	C(C)— Contamination- Chemical	Possible	Limited	Slow	0.8
Water	Chlorination	AT1—Assault Team 1	Possible	Limited	Slow	0.8
Water	Distribution Tank	AT1—Assault Team 1	Possible	Limited	Slow	0.8
Wastewater	Lift Stations	C(C)— Contamination- Chemical	Possible	Limited	Slow	0.8
Wastewater	Lift Stations	F1—100-year flood	Low	Low	Low	1.0
Water	Pump Station	F1—100-year flood	Moderate	Moderate	Moderate	0.72
Water	SCADA	(C1)— Cyberattack	Possible	Limited	Slow	0.8
Wastewater	SCADA	(C1)— Cyberattack	Possible	limited	Slow	0.8

5.2.2.5   Step 5—threat analysis.   The purpose of the threat analysis is to answer the question: "What is the likelihood that man-made hazards (malevolent, natural hazard, or dependency/proximity hazard) will strike the utility's facility?"

Table 5-15 Step 5–threat analysis

J100-21 Step Requirements		Case Study Application
4.5.1	Malevolent threats. Malevolent actors, domestic or international terrorists, disgruntled insiders or customers, and criminals pose man-made threats to critical infrastructure. The principal focus of this standard is on threat actors whose actions would likely be classified as terrorism. A terrorist is typically motivated to kill or injure as many people as possible and/or to cause massive economic and social disruption to spread fear. They may be willing to lose their lives to make their attacks succeed. Their human resources range from a single assailant to a team of a dozen or more, and their material resources range from simple firearms to explosives, contaminants, and/ or "weaponized" vehicles, e.g., the airplanes in the World Trade Center attack on Sept. 11, 2001, or the truck in the Nice, France, promenade on July 14, 2016.	The threat likelihood method described in Sec. 4.5.2.1 (with details in Sec. 4.5.2.1.1 through 4.5.2.1.6) in the standard was used, which resulted in a man-made threat likelihood as shown in Table 5-16 Malevolent Threat Likelihood Calculations using Proxy Method.

Table 5-15   Step 5–threat analysis (*Continued*)

J100-21 Step Requirements	Case Study Application
This standard also recognizes the threats associated with criminals and disgruntled employees or customers. Information supporting assessment of these threats is based on the utility's and similar utilities' experience, including measures taken to mitigate persistent acts that result in disruption of services such as theft of materials and resources. If the action of these threat actors escalates, for purposes of the standard, such an incident would be addressed as reference threats associated with sabotage or terrorist attack(s).	
While a definitive database or expert source of terrorist threat likelihoods down to the threat-asset level would be preferred, none exists today and is unlikely in the foreseeable future. If a definitive, authoritative source or method becomes available, it should be used. Such authoritative sources for terrorism likelihood guidance might include recognized federal, state, and local intelligence agencies; homeland security; and/or law enforcement agencies.	
There are two approaches for estimating malevolent threat likelihood: (1) The "proxy" method (Sec. 4.5.2) developed by this standards committee and (2) Best Estimate (Sec. 4.5.3). Each is discussed subsequently.	
4.5.2 Proxy method.	The detail requirements are not repeated here, but the analysis for this case study uses this method. The calculations are shown in Table 5-16 Malevolent Threat Likelihood Calculations Using Proxy Method.
4.5.3 Best estimate. Using this method, likelihood is determined based on informed experience of the organization and input from federal, state, and local law enforcement and others. The likelihood will be a probability with a value between 0.0 and 1.0. USEPA has developed a resource entitled "Baseline Information on Malevolent Acts for Community Water Systems" that may be used to inform the best estimate approach. The utility shall document and record all assumptions, data sources, and reasoning used in estimating terrorism threat likelihood.	This case study used the proxy method for malevolent threats.
4.5.4 Cyber threats. Given the prevalence and expanding threat surface associated with cyberattacks on enterprise and process controls systems, it is recommended that water systems use 1 as the threat likelihood for cyber. National intelligence agencies have consistently stated that cyber is the top threat facing critical infrastructure and business. The threats are agnostic to size and location, and there is no rational justification to believe that attempts to attack a water utility, actively or passively, will never occur or become less frequent given growth of technology applications. The effect of this recommendation is that every utility should expect that they will be attacked in some manner and that appropriate cyber risk management control should be implemented as appropriate consistent with their technology applications.	The threat likelihood for cyber threats in this analysis is assumed to be 1.0.

(continued)

Table 5-15   Step 5–threat analysis (*Continued*)

	J100-21 Step Requirements	Case Study Application
4.5.5	Natural hazards. Estimate the probability of natural hazards by averaging the historical record of frequencies for each intensity or magnitude of the hazard and the specific location of the asset. Federal and state agencies collect and publish data for hurricanes, earthquakes, tornadoes, wildfires, and floods, which can be accessed for actual frequencies for various levels of severity of natural hazards. The specific sources and methods are provided in the J100 Operational Guidance. Preference should favor the more local data to broader aggregates (e.g., county level preferred to state level) when available. These may be adjusted slightly where trends in the data suggest use of historical averages would err relative to future conditions. Moving averages would identify where this approach is needed. Consultation with hazard-specific subject matter experts at federal or state agencies or universities may be considered as well. Document in detail all sources, assumptions, and any adjustments or calculations used to estimate the likelihood of natural hazards.	The natural hazard threat that is included in the analysis is the 100-year flood. By definition, it has a likelihood of 1 out of 100 or 0.01 per year. If other natural threats were to be considered, then historical data should be used for the likelihood.
4.5.6	Dependency and proximity hazards. An initial basis for estimates of the likelihood, severity, and duration of service outages or denials by organizations that supply the utility are based on local historical records for critical suppliers to the utility. These estimates may serve as a baseline estimate of "business as usual" or incrementally increased if the analysis indicates the value may be higher due to malevolent or natural hazard activity on the required supply chain elements. These suppliers may include both physical (e.g., source water, chemicals, electricity, fuels, parts, and equipment) and cyber (e.g., process controls and business processes). Confidential, highly candid conversations with the actual suppliers may provide guidance on whether the historical experience is likely to continue or change in specific ways. The geographic location of suppliers' assets may affect which of the utility's assets would be affected. Asking the supplier to estimate service continuity and/or the likelihood of service interruption at specific locations of the utility's critical assets could provide invaluable insight into the suppliers' resilience. Hazards that would affect both the utility and its suppliers (e.g., hurricanes or earthquakes) should be discussed in detail.

Estimate the likelihood of incurring collateral damage from an attack, failure, or natural hazard on a nearby or co-located asset based on the local situation and by using the same logic in estimating malevolent threats (Sec. 4.5.1). | None included in this analysis. |
| 4.5.7 | Record the methods and estimates for threat likelihoods. Record the method used for making the estimates in detail and the estimates themselves, as either single-valued point estimates, ranges, or probability distributions. | Shown subsequently in Table 5-16 for man-made threats. |

Table 5-16   Malevolent threat likelihood calculations using proxy method

	Description	AT1	C(C) Chemical	Remarks
Nation	Probability of attack in United States	0.002	0.002	Per standard, used default value percentage of threats to United States from world population
Metro region	Likelihood of metro region	0.002	0.002	Using value for "all other" since our location is not listed in standard table
Broad target type	Water utility	0.0769	0.0769	Using category 7, 2/36 per standard J100-21, Table 4
Specific utility	Per Rand table	0.1	0.1	Estimated that 10% of utility service population served
Adjust for detection and interdiction	Per table	0.5	0.15	Using .5 interdiction for AT1 and .85 for contamination
Select a threat–asset pair	Asset likelihood	0.9	0.9	Assumed that all assets are attractive
	Proxy threat likelihood calculations	**1.38E-08**	**4.155E-09**	

5.2.2.6   Step 6—Risk/resilience analysis.   The purpose of the risk/resilience analysis step is to answer the questions: "What is my risk? What is my resilience?" The risk analysis uses the results from the consequence, vulnerability, and threat evaluations.

Table 5-17   Step 6—risk/resilience analysis

	J100-21 Step Requirements	Case Study Application
4.6.1	Estimate risk for each threat-asset pair as the function of Consequence Analysis (see Sec. 4.3), Vulnerability Analysis (see Sec. 4.4), and Threat Analysis (see Sec. 4.5) using the equation: Risk = Consequences × Vulnerability × Threat Likelihood = $C \times V \times T$	Risks were calculated using the results of earlier steps, and results are shown in Table 5-18. Note that it was decided to show the "monetized risk" values.
4.6.2	Per J100-21 4.6.1, risk is to be calculated for EACH term: (1) human casualties (life equivalents), (2) service outage (gallons), (3) financial loss to the owner (dollars), and (4) regional economic impacts (dollars).	
4.6.3	Display estimated fatalities and injuries. When displaying the utility's risk of financial loss or the community's loss, the utility shall also display the estimated number of fatalities and serious injuries weighted by the threat likelihood and vulnerability for the analysts and decision-makers.	Estimated fatalities and injuries are also shown in Table 5-18.
4.6.4	Assess the overall utility resilience by estimating the Utility Resilience Index (URI).	The URI was estimated using the method specified in the standard and is shown in Table 5-18.

Table 5-18   Risk analysis results

System	Asset	Threat	Number of Fatalities	Number of Injuries	Fatalities ($)	Injuries ($)	Utility Financial Impact	Regional Economic Impact	Detection / Preparation	Delay/ Active Response	Response/ Recovery	Likeli-hood of Damage	Threat Likeli-hood	Risk — Calculated from Consequence x Vulnerability x Threat Likelihood
Water	Chlorination	F1—100-year flood	0	0	$0	$0	$40,000	$50,000	Moderate	Moderate	Moderate	0.72	0.01	$648
Water	Distribution Tank	C(C)—Contamination-Chemical	0	0	$0	$0	$50,000	$0	Possible	Limited	Slow	0.8	4.15385E-09	$0
Water	Chlorination	AT1—Assault Team 1	1	15	$7,800,000	$12,285,000	$100,000	$0	Possible	Limited	Slow	0.8	1.38462E-08	$0
Water	Distribution Tank	AT1—Assault Team 1	0	0	$0	$0	$60,000	$0	Possible	Limited	Slow	0.8	1.38462E-08	$0
Wastewater	Lift Stations	C(C)—Contamination-Chemical	0	0	$0	$0	$200,000	$0	Possible	Limited	Slow	0.8	4.15385E-09	$0
Wastewater	Lift Stations	F1—100-year Flood	0	20	$0	$16,380,000	$50,000	$100,000	Low	Low	Low	1	0.01	$165,300
Water	Pump Station	F1—100-year Flood	0	0	$0	$0	$40,000	$50,000	Moderate	Moderate	Moderate	0.72	0.01	$648
Water	SCADA	(C1)—Cyberattack	0	0	$0	$0	$80,000	$100,000	Possible	Limited	Slow	0.8	1	$144,000
Wastewater	SCADA	(C1)—Cyberattack	0	0	$0	$0	$60,000	$100,000	Possible	limited	Slow	0.8	1	$128,000

Table 5-19   Utility Resilience Index for case study

J	URI Indicators	Utility Profile	Weight	Value	Weight x Value
O1	**Emergency Response Plan (ERP)**			0.139	0.0695
	No ERP		0		
	ERP developed and/or updated		0.25		
	Staff trained on ERP (i.e., Table Top)	X	0.5		
	Resource typed assets/teams defined and inventoried		0.75		
	Functional exercises on the ERP conducted		1		
O2	**National Incident Management System (NIMS) Compliance**			0.156	0.078
	No ICS/NIMS Training		0		
	ICS 100/200 provided to key staff		0.25		
	ICS 700/800 provided to key staff	X	0.5		
	ICS 200/300 provided to key staff		0.75		
	Utility certified as NIMS compliant		1		
O3	**Mutual aid and assistance**			0.187	0.14025
	None		0		
	Mutual Aid/Intermunicipal (within own city/town agencies)		0.25		
	Mutual Aid/Local-Local (with adjacent city/town)		0.5		
	Mutual Aid/Intrastate (e.g., Water/ Wastewater Agency Response Network [WARN])	X	0.75		
	Mutual Aid/Interstate and Intrastate		1		
O4	**Emergency power for critical operations**			0.06	0.03
	None		0		
	Up to 24 h		0.25		
	25–48 h	X	0.5		
	49–72 h		0.75		
	Greater than or equal to 73 h		1		
O5	**Ability to meet minimum daily demand (water) or treatment (wastewater)**			0.097	0.0485
	None		0		
	Up to 24 h		0.25		
	25–48 h	X	0.5		
	49–72 h		0.75		
	Greater than or equal to 73 h		1		

(continued)

Table 5-19   Utility Resilience Index for case study (*Continued*)

J	URI Indicators	Utility Profile	Weight	Value	Weight x Value
O6	**Critical parts and equipment**			0.088	0.088
	3–4 weeks or greater		0		
	1–<3 weeks		0.25		
	3–<7 days	X	0.5		
	1–<3 days		0.75		
	Less than 24 h		1		
O7	**Critical Staff Resilience**			0.06	0.03
	<10%		0		
	10–25%		0.25		
	>25–50%	X	0.5		
	>50–75%		0.75		
	>75–100%		1		
F1	**Business Continuity Plan**			0.046	0.023
	No BCP		0		
	BCP under development		0.25		
	BCP completed	X	0.5		
	BCP fully implemented		0.75		
	Annual commitment of resources and BCP exercised		1		
F2	**Utility Bond Rating**			0.064	0.048
	Caa, less than or equal to		0		
	B-Ba		0.25		
	Baa-A		0.5		
	AA	X	0.75		
	AAA		1		
F3	**GASB Assessment**			0.017	0.01275
	Less than 20% assessed		0		
	21–40% assessed		0.25		
	41–60 % assessed		0.5		
	61–80% assessed	X	0.75		
	Greater than 81% assessed		1		
F4	**Median household income**			0.046	0.023
	≥ 5% National Average		0		
	> 2–4% National Average		0.25		
	+/–2% National Average	X	0.5		
	< 2–4% National Average		0.75		
	≤ 5% National Average		1		

(*continued*)

Table 5-19   Utility Resilience Index for case study (*Continued*)

J	URI Indicators	Utility Profile	Weight	Value	Weight x Value
F5	**Unemployment**			**0.04**	**0.02**
	≥ 10% National Average		**0**		
	> 5–10% National Average		**0.25**		
	+/–5% National Average	**X**	**0.5**		
	< 5–10% National Average		**0.75**		
	≤ 5% National Average		**1**		
			Total URI Calculated		**0.611**

5.2.2.7   Step 7—Risk/resilience management.   The purpose of the risk/resilience management step is to answer the questions: "What options does the utility have to reduce risks and increase resilience? How much will each benefit in reduced risks and increased resilience? How much will it cost? What is the benefit–cost ratio of the utility's options?"

Table 5-20   Step 7–risk and resilience management

	J100-21 Step Requirements	Case Study Application
4.7.1	Decide what risk and resilience levels are acceptable by examining the estimated results of the first six steps for each asset–threat pair. For those that are acceptable, document the decision. For those that are not acceptable, proceed to the next steps.	Table 5-17 has the details of the risk calculation for this case study. A summary of the risks calculated is in Table 5-20. The highest risk calculated is from a flood of the wastewater lift station, followed by cyberattacks on the water and wastewater SCADA systems. The other threats considered have a much lower risk calculated, primarily because the threat likelihood proxy method was used for the man-made threats. The results would be different with an alternate threat likelihood calculation method. In this case study, management has decided they want to address the top three risks and the other risks may be addressed at a later time.
4.7.2	Define risk management options for the risks that are unacceptable to the utility. Consider the alternatives of simply accepting the risk, transferring the risk, and nontransferable options as alternatives in evaluating which may be most effective and efficient. These might include countermeasures and mitigation/resilience measures, mutual assistance programs, emergency responses and rapid recovery, and risk transfer options. Insurance and other forms of risk transfer should be negotiated with appropriate brokers, insurers, and re-insurers to determine whether risk transfer insurance can be obtained at costs less than the estimated risk. As an alternative to risk transfer or where utilities are self-insured, define options to reduce risks and/or enhance resilience. Specifically indicate which term(s) of the risk and resilience equations—threat likelihood, vulnerability, specific consequences including service outages—would change, by how much due to each option, and why.	Potential countermeasures for reducing risk from the highest asset–threat pairs are 1) build a flood barrier around the wastewater lift station above the 100-year flood level (termed CM2) and 2) institute a cybersecurity program that would benefit both the water and wastewater SCADA assets (termed CM3); see Table 5-23.

Table 5-20   Step 7–risk and resilience management (*Continued*)

	J100-21 Step Requirements	Case Study Application
4.7.3	Estimate the life-cycle costs needed to implement and operate the option. Life-cycle costs include investment and operating costs of each option, regular maintenance, and periodic overhaul, if expected, human time and effort, contract support, etc. If costs will be incurred over more than one year, the full stream of costs must be converted to present value. Also estimate the first budget period outlay for budgeting.	Estimated resources are shown in Table 5-21. The engineer of record provided preliminary cost estimates. To support decision-making, the costs are shown as life-cycle costs with the assumptions shown in Table 5-23.
4.7.4	Assess the options by analyzing the threat–asset pair under the assumption that the option has been implemented—revisiting all affected steps in Sec. 4.3 through Sec. 4.6 to re-estimate each respective risk (human casualties, service outage, financial loss to the owner, and economic loss to the region) and the composite risks to the utility and the community's well-being. Then calculate the estimated gross benefits of the option. Gross benefits are the difference between the risk levels without the option and those with the option in place. If the benefits are incurred over more than one year, they should be converted to present values.	The countermeasure CM2 would reduce the consequences to wastewater lift station to the 100-year flood, and the resulting risk to zero. The calculations of risk reduction are in Table 5-22. The countermeasure CM3 would reduce the vulnerability to the cyber threats substantially. The threat likelihood would not change, but the vulnerability would change from 80% to 20%.
4.7.5	Identify the options that have benefits that apply to multiple threat–asset pairs. For example, if a higher fence changes the vulnerability for an attack by one assailant as well as an attack by two to four, the gross benefits of the two threat–asset pairs should be added together as the benefit of the combined option. Similarly, if an option reduces the vulnerability to threat–asset pairs in addition to the one(s) for which it was initially designed, e.g., the higher fence would reduce vulnerability of many assets in the fenced area. For some options, there may be many affected assets; all should be included in estimating gross benefits. At times, introduction of a broad risk-reduction option that lowers risks for several threat–asset pairs may reduce the benefits of other options taken individually (e.g., the benefits of a hardened door will be different based on whether an effective security fence around the asset is in place or not.) These circumstances should be identified and analyzed carefully to configure the options as portfolios that make sense to decision-makers.	See discussion in Sec. 4.7.1–4.7.4 in the left column.
4.7.6	Calculate the net benefits (gross benefits minus life-cycle costs). Net benefits are the measure of the *value* of the option, and the benefit/cost ratio (and/or other criteria that are relevant in the utility's resource decision-making, e.g., return on investment), and the measure of the option's efficiency in producing benefits per dollar of cost.	The measure chosen for evaluation is annual dollars per risk reduction and is in Table 5-23.
4.7.7	Review and select the options considering at least • Human casualties benefits, • Mission continuity (resilience) benefits (reduction of outage risks), • Utility's financial benefits, • Utility's combined benefits, • The community's gross regional product benefits, and • The combined regional well-being benefits.	This is a management action to review the results presented.

(continued)

Table 5-20 Step 7–risk and resilience management (*Continued*)

J100-21 Step Requirements	Case Study Application
Start this consideration by ranking the options by net benefit for each type of benefit. Individual qualitative benefits (avoidance of damage to reputation, environmental damage avoided, etc.) should also be displayed for the options as ranked by each quantitative benefit. The qualitative benefits are not used in the rankings but may enter into the decisions as final selections are made. Tentatively select the options with the greatest net benefit, using benefit/cost ratio as "tie-breaker" of options with very similar levels of net benefits. Selecting based on net benefits is consistent with the objective of maximizing net benefits. Some organizations use the benefit/cost ratio to guide selection, but this practice can result in a lower level of total benefits. Benefit/Cost ratio is a measure of efficiency—benefits per dollar—which will only maximize benefits if there are virtually no constraints (e.g., available budget). Where constraints apply, this can result in a portfolio of the most efficient options, but not the greatest total amount of net benefits. The same is true for return on investment. Display this information for the decision-makers in a format with which they are familiar.	
4.7.8 Document the decision process and the resulting choices, including all assumptions, outside data, logic of the choice process, etc. Make clear the expected net benefits from each selected option for use in progress monitoring and evaluation.	Any countermeasures taken credit for in the analysis and added should be regularly reviewed and tested to make sure that they are serving the intended function. This is a practice recommended by multiple organizations.

Table 5-21 High-level risk-assessment summary

System	Asset	Threat	Threat Likelihood	Risk Calculated from Consequence × Vulnerability × Threat Likelihood
Water	Chlorination	F1—100-year Flood	0.01	$648
Water	Distribution Tank	C(C)—Contamination-Chemical	4.15E-09	$0
Water	Chlorination	AT1—Assault Team 1	1.38E-08	$0
Water	Distribution Tank	AT1—Assault Team 1	1.38E-08	$0
Wastewater	Lift Stations	C(C)—Contamination-Chemical	4.15E-09	$0
Wastewater	Lift Stations	F1—100-year flood	0.01	$165,300
Water	Pump Station	F1—100-year flood	0.01	$648
Water	SCADA	(C1)—Cyberattack	1.00	$144,000
Wastewater	SCADA	(C1)—Cyberattack	1.00	$128,000

The countermeasures discussed subsequently are only a representative sample of risk countermeasures for this sample utility. When selecting countermeasures to carry forward in the risk assessment (cost-benefit analysis), the countermeasures' ability to reduce and/or mitigate the risk of asset–threat pairs should be evaluated and thus prioritized. Countermeasures may include planning and procedural improvements, capital projects, or operational modifications at the utility's facilities. When brainstorming risk countermeasures, impact on the utility's risk and overall project timeline need to be considered along with how the countermeasure may be funded and impact operations and maintenance. The utility's existing capital improvement plan should be reviewed during this step because some risk countermeasures may already be part of a future utility project. Countermeasures may not be limited to only capital infrastructure projects. Preparing the utility's staff to respond to an emergency through tabletop exercises and other formal training can be a cost-effective way of mitigating risk.

When developing countermeasures, consideration should be given to the public's welfare and the utility's staff's welfare.

- CM1 includes replacing gaseous chlorine with sodium hypochlorite. This countermeasure aims to eliminate the threat of accidental or intentional release of gaseous chlorine. Gaseous chlorine can pose a significant threat to the utility's staff and the community surrounding the facility where the gaseous chlorine is used.

- Consideration should be given to developing countermeasures that reduce the duration of service loss since the utility is failing to maximize its revenue when it is unable to deliver water to its customers. One example countermeasure relating to service loss is installation of standby generators at water treatment facilities in the event power is lost at the facility.

- An example relating to service disruption and power loss countermeasure is provision of a redundant power feed line at mission critical facilities that only have one source of power. An example relating to infrastructure is building a redundant, parallel transmission main in the event a line break occurs that would have normally taken out service to an entire portion of the utility's service area.

Table 5-22   Risk reduction from countermeasure additions

System	Asset	Threat	Original Risk Calculated from Consequence × Vulnerabilty × Threat Likelihood	Risk After Adding CMs Calculated from Consequence × Vulnerabilty × Threat Likelihood	Risk Decrease Amount Risk Reduced by Addition of CMs	CM Applied
Water	Chlorination	F1—100-year flood	$648	$288	$360	CM1
Water	Distribution Tank	C(C)—Contamination-Chemical	$0	$0	$0	
Water	Chlorination	AT1—Assault Team 1	$0	$0	$0	CM1
Water	Distribution Tank	AT1—Assault Team 1	$0	$0	$0	
Wastewater	Lift Stations	C(C)—Contamination-Chemical	$0	$0	$0	
Wastewater	Lift Stations	F1—100-year Flood	$165,300	$0	$165,300	CM2
Water	Pump Station	F1—100-year Flood	$648	$648	$0	
Water	SCADA	S(CU)—Hacker	$144,000	$36,000	$108,000	CM3
Wastewater	SCADA	S(CU)—Hacker	$128,000	$32,000	$96,000	CM3

Table 5-23   Countermeasure costs and benefits

		Capital Costs (Note 1)	Operating Costs/Year (Note 2)	Amortization Period (Years) (Note 3)	Annual Present Value	Eliminated Costs/Year	Annualized Cost	Risk Reduction	Risk Reduction ($)/ Annualized Cost ($)
CM1	Replace gaseous chlorine with sodium hypochlorite	$115,000	$30,000	20	$35,750	$10,000	$25,750	$360	0.01
CM2	Build flood barrier around wastewater lift station	$65,000	$1,000	20	$4,250	$0	$4,250	$165,300	38.89
CM3	IT improvement package for water and wastewater SCADA	$120,000	$30,000	15	$42,000	$0	$42,000	$204,000	4.86

NOTES:
1: Includes all engineering, permitting, and construction costs
2: Includes expected supplies and maintenance costs
3: Assumed life for amortization
4: Assumed no interest in calculating annual present value, capital cost divided by amortization period plus operating costs per year
5: CM1 eliminated cost is the cost of gaseous chlorine and periodic Risk Management Plan RMP; no eliminated costs for CM2.

APPENDIX A

Contamination

SECTION A.1:   INTRODUCTION

Contamination of the water supply is possible at many points of the production and distribution processes and can occur from natural events, accidental spills and leaks, or malevolent acts. It is necessary to evaluate where a contaminant could be introduced (the asset); the possible contaminants, methods of contamination, and responsible parties (the threat); how readily the asset can be contaminated (the vulnerability); and the resulting aesthetic, operational, financial, reputational, and health effects (the consequences).

SECTION A.2:   LOCATION OF CONTAMINATION

Many utilities do not own or control the entire watershed of surface water sources or the contributory radius to groundwater sources, and even with utility control, these areas can be difficult to fully protect. Contamination of source water can occur underground with plumes of pollution, such as those from a leaking underground storage tank; over land by runoff from contaminated sites; or by air with dispersion of a contaminant, either naturally or via mechanical distribution methods such as a crop duster. Contamination can also occur within the treatment facility, such as by overfeed or spill of chemicals, cross-connection, or material incompatibility. Contamination of finished water is possible in finished water storage facilities, in the distribution system through customer cross-connections, or through leaking pipes or loose joints combined with low-pressure events. Any point of the process could be vulnerable to contamination by intentional tampering, either by directly introducing a contaminant at the unit process or by manipulating control systems to deliberately

53

cause one of the previously listed events to occur. For purposes of this operational guide, this section will focus on risks to source water.

SECTION A.3:   TYPES OF CONTAMINANTS

Several contaminants present a risk to water utilities through both natural and man-made pathways. Resources are available from the US Environmental Protection Agency (USEPA) with detailed lists of contaminants, and they are described generally as follows.

Biotoxins.   Biotoxins are generally thought to include toxins that are deliberately introduced by a malevolent actor as part of a terrorist attack. However, this category may also include naturally occurring substances such as cyanotoxins, which are produced naturally by cyanobacteria in source water and can have immediate health effects on children and animals.

Chemicals.   Like biotoxins, chemicals are a particular risk if they are not part of normal testing by the water utility as this can delay detection. Chemical contamination can present risk of immediate illness or chronic illness. Chemical risks can include:

- chemicals used or stored by the utility itself, or related functions that may share spaces with the utility (e.g., public works facilities), such as:
  - water treatment chemicals,
  - fuel,
  - chemicals used for other functions, such as boiler water treatment, and
  - products used for maintenance such as paints, lubricants, deicers, and cleaning solutions;
- chemicals used or stored by others on properties within the zone of influence of the source water, either above or below ground;
- chemicals for which bulk delivery vehicles such as trucks and freight trains regularly travel near water sources;
- chemicals used for emergency response, such as firefighting foam and adsorbents;

- chemicals used or stored by customers that could be introduced into the distribution system via a cross-connection; and
- any chemicals that could be introduced as part of a malevolent attack.

Explosives. Explosives can lead to water contamination by causing physical damage to storage, treatment, and distribution facilities, which can allow contamination to occur naturally or deliberately.

Pathogens. Pathogen contamination is likely the most common type of drinking water contamination. Pathogen contamination sources can include:

- runoff from sites frequented by livestock or wildlife,
- leaking, damaged, or improperly designed or managed sewage and septic systems,
- flooding,
- sewer overflows,
- cross-connections with wastewater treatment facilities and other nonpotable uses, such as irrigation water, and
- animals, insects, and birds accessing unsecured finished water storage tanks.

Radionuclides. Radionuclide contamination of drinking water typically originates from natural elements in adjacent soil but can also occur from improper storage and disposal of radioactive materials. The likelihood of radionuclide presence can often be determined by geographical proximity to areas of known radionuclide deposits and registered users of radionuclide materials. Since radionuclide monitoring is performed infrequently for water utilities with a history of low or nondetectable levels of radionuclides, a new instance of radionuclide contamination might not be discovered for several years.

SECTION A.4: CONSEQUENCES

The consequences to the facility owner from drinking water contamination are the sum of the individual losses. These may include:

Loss of life and serious injuries. The most severe consequence of drinking water contamination is illnesses and fatalities among those who consume or come in contact with the water, which can occur in a short timeframe following the contamination event, can be widespread, and might not be immediately suspected to be originating from the water supply. The severity of this consequence will vary by the number of people potentially affected, the duration of the event and the toxicology of the contaminant, and whether health effects are acute or chronic. For purposes of the risk calculation, convert potential losses to dollars if utilizing the J100-21 Statistical Values of Life and Serious Injury in the assessment.

Service denial. Lost income due to the time required to remediate the contaminated asset: This is the dollar value of income lost (time of service x MGD lost x $/MGD) while the asset is unusable for cleaning and repair of vulnerabilities that allowed the contamination.

Owner's financial loss. This includes:

- One of the major consequences of any contamination event is loss of confidence in the water system as a supplier of drinking water. This can happen even if the chemical introduced is not actually harmful. There is a psychological impact on the users, a potential financial impact on the users in buying bottled water, and a financial impact on the utility from lost revenue due to customers using less tap water;

- lost income due to disposal of contaminated treated water (MG lost × $/MG);

- cost of additional treatment needed to address contamination (time treatment is needed × MGD × $/MGD);

- cost of providing alternate water supplies while water supply is not suitable for all uses ($/MGD);

- cost of performing additional water quality analyses to identify type, extent, and duration of contamination;

- cost of public outreach needed to issue and lift associated advisories; and

- potential financial impact if utility is found to be legally liable for contamination impacts.

Economic losses. This applies to the regional economy for an extended outage.

Consequences can be estimated using known costs from everyday operations, case studies of similar actual incidents, and guidance provided in J100-21 and other resources listed at the end of this appendix.

SECTION A.5: VULNERABILITY

The vulnerability of a water utility to contamination is the conditional likelihood of the worst reasonable case happening given that the threat occurs. J100-21 defines vulnerability estimate as the conditional probability that a specific threat, given that it occurs, will cause the specific estimated consequence to the asset. The following questions may be used to guide this evaluation. Any value between 0 and 1 may be assigned. Selected points are provided to help "anchor" the respective vulnerability scales. They are not to be assumed as the only values to be selected; interpolation is expected to give the analyst the full range to work with.

1. Is the source water vulnerable to contamination?
   - 0: If the source is completely secure and/or is sufficiently large in volume to dilute any probable type of contamination sufficiently to prevent adverse impact
   - 0.5: If the source has some security vulnerabilities and/or is small enough in volume that contamination is likely to require additional treatment or source supplementation
   - 1: If the source cannot be fully secured (open surface water source) and/or contamination will result in loss of the source

2. Do sources of contamination exist near the source?
   - 0: If the source, watershed, and/or wellhead area is owned and controlled by the water utility and well secured
   - 0.5: If nearby land uses present contamination risks but are minor, documented, and/or controlled
   - 1.0: If nearby sources of contamination are known or suspected but not subject to proper controls

3. Is there a significant risk of malevolent tampering with the treatment process?

57

- 0: Strict access controls and a robust cybersecurity program are in place to prevent and rapidly detect unauthorized access to treatment and control systems and chemicals.
- 0.5: Physical security and cybersecurity measures are in place but would not completely prevent access to treatment and control systems and chemicals.
- 1: Physical security and cybersecurity are minimal and would not prevent access to treatment and control systems and chemicals.

4. Is there a risk of accidental contamination within the treatment facilities?
   - 0: No items presenting contamination risks are used at or near the facility; robust internal cross-connection control program is in place.
   - 0.5: Materials and processes that could present contamination risks are present but monitored, and employees are trained and audited in preventive measures.
   - 1.0: Multiple possible contamination risks are present in or near facilities with little monitoring or training.

5. Is there a significant risk of contamination of treated water?
   - 0: Strong physical security on all finished water storage and access points; a robust cross-connection control program is in place and is likely to prevent contamination.
   - 0.5: Points of physical vulnerability exist including finished water storage and access points, and risk of cross-connections.
   - 1: Little or no cross-connection control program; severe physical security deficiencies in finished water storage.

6. Do processes exist to rapidly detect distribution system water quality changes?
   - 0: Online water quality monitoring technology present throughout system; frequent communication with health officials.
   - 0.5: Distribution system sampling conducted frequently in response to customer complaints and observations of quality changes (odor, color).

- 1: Distribution system monitoring is conducted only as required by applicable regulations.

For major contamination risks, vulnerability may be estimated using the more sophisticated approaches described in the standard (e.g., event trees, fault trees). These would break down the overall vulnerability into a series of individual elements that may be easier for analysts to judge.

SECTION A.6:   THREAT LIKELIHOOD

The likelihood of a contamination event can be difficult to estimate as it highly depends on multiple site-specific factors besides the vulnerability. Accidental contamination likelihood depends on the proximity of contaminants to unprotected facilities and how these contaminants are handled and stored. Deliberate contamination likelihood depends on the motivation and intent of the actor.

One resource to estimate a numeric value for risk calculation is presented in the USEPA document "Baseline Information on Malevolent Acts for Community Water Systems v. 2.0" (February 2021). This document provides the following default threat likelihoods for contamination:

- Contamination of Source Water—Accidental: 0.05
- Contamination of Source Water—Intentional: 10^{-6}
- Contamination of Finished Water—Accidental: 0.2
- Contamination of Finished Water—Intentional 10^{-5}

The guidance provides a series of questions for each of the listed scenarios to enable modification of the default threat likelihood for a specific utility for its own risk assessment.

SECTION A.7:   RISK CALCULATION

The total risk (R) of an asset that is affected by contamination is the product of the consequences, vulnerability, and threat likelihood.

$$R = C \times V \times T$$

Example:

Asset: 1 mil gal treated water storage tank

Location: Denver, Colo.

Threat: Accidental contamination from wildlife entering unsecured tank

Consequences

Cost to drain and clean tank:	$400,000
Lost revenues/loss of service:	1 MG × 2.50/gal initial supply = $2,500,000
Estimated illnesses:	50

Injury consequence calculated at 50 injuries valued at 0.25 × statistical value of life $7.6 million: $95,000,000

Total consequence for this threat: $400,000 + $2,500,000 + $95,000,000 = $97,900,000

Vulnerability

Conditional likelihood of damage given the contamination threat occurs, resulting in contamination of stored water (using 1 C above) = 1.0

Likelihood

Likelihood of the threat contamination of finished water, accidental threat, per USEPA guidance = 0.2

Risk Calculation

Risk due to this threat = $C \times V \times L = \$97,900,000 \times 1.0 \times 0.2 = \$19,580,000$

SECTION A.8:   RESOURCES

National Research Council. 2006. *Drinking Water Distribution Systems: Assessing and Reducing Risks*. Washington, DC: The National Academies Press.

Seidel, C., A. Ghosh, G. Tang, S. Hubbs, R. Raucher, and D. Crawford-Brown. 2014. "Identifying Meaningful Opportunities for

Drinking Water Health Risk Reduction in the United States." Water Research Foundation Report 4310 and RHI Calculator tool. Denver: Water Research Foundation.

US Environmental Protection Agency (USEPA), Office of Ground Water and Drinking Water. 2003. "Response Protocol Toolbox: Planning for and Responding to Drinking Water Contamination Threats and Incidents." Overview (EPA-817-D-03-007); Water Utilities Planning Guide—Module 1 (EPA-817-D-03-001); Contamination Threat Management Guide—Module 2 (EPA-817-D-03-002); Site Characterization and Sampling Guide—Module 3 (EPA-817-D-03-003); Analytical Guide—Module 4 (EPA- 817-D-03-004); Public Health Response Guide—Module 5 (EPA-817-D-03-005); and Remediation and Recovery Guide—Module 6 (EPA-817-D-03-006).

USEPA. 2003. *Cross-Connection Control Manual.* EPA 816-R-03-002. Washington, DC: USEPA.

USEPA. 2021. "Baseline Information on Malevolent Acts for Community Water Systems v. 2.0." EPA 817-F-21-004.

USEPA. 2021. "Occurrence of Releases with the Potential to Impact Sources of Drinking Water." EPA 817-R-21-001.

World Health Organization. 2004. "Chemical Safety of Drinking Water: Assessing Priorities for Risk Management." Geneva, Switzerland: World Health Organization.

This page intentionally blank.

APPENDIX B
Cyber

SECTION B.1: INTRODUCTION

Cyber threats are occurring more frequently each year and have larger consequences. Information technology (IT) networks and process control system (PCS) networks are targeted and successfully compromised with financial losses and denial of service impacts.

The approach discussed in the operational guide allows the user to use the J100-21 methodology to assess cyber, physical, and natural hazard risks and risk-reduction options in direct, comparable terms. Under this approach, the same concepts, risk formula, and key terms (consequence, vulnerability, and likelihood) are used to assess risk for physical threats, natural hazards, and dependency hazards:

Risk = Consequence × Vulnerability × Threat Likelihood = C × V × T

- *Consequence* is measured in dollars of financial loss to the utility caused by damage, denial of service, human casualties, and regional impacts.
- *Vulnerability* is the conditional probability that the estimated consequences will result from the threat incident, given the incident occurs.
- *Threat likelihood* is the probability or frequency [see later discussion] that the incident will happen in the next 12 months, measured in occurrences per year.

The risks of interest in J100-21 are the expected value (i.e., weighted probability) of the dollars of loss to the utility, expected gallons of service interruption, expected life-equivalent casualties, and expected regional economic impacts. Risk is measured in average dollars/year.

Some organizations consider Impact = Consequence × Vulnerability.

Other organizations consider Impact = Threat Likelihood × Vulnerability.

Limitations of This Approach

Physical attacks or natural hazards to the physical IT or PCS infrastructure are covered in other areas. For example, a hurricane destroys the building with the data center holding the IT servers. The consequence of the loss of the servers, the cost to purchase new ones, and rebuilding them from backups are part of the consequences of the threat–asset pair of the hurricane (threat) and the building (asset).

Cyber Assets

Risk is calculated for each threat–asset pair. Asset selection and threat selection are important to "right size" the cyber assessment.

The water/wastewater department may be a portion of a larger organization. The organization may have departments for police, courts, jail, fire, libraries, streets, airport, parks, and recreation just to name a few possibilities. There are cyber threats that may impact the entire organization including water/wastewater. The assessors need to determine the water/wastewater portion of the greater organization. The proportions can be based on any reasonable division. Examples may include the number of employees in water/wastewater or number of servers and computers for the water/wastewater department or any other reasonable division. For example, if ransomware is released on the IT network and all enterprise computers are encrypted, the water/wastewater department should contribute a percentage of the total recovery cost based on their portion of use of shared assets. This can be based on the ratio of water/wastewater employees to total organization employees or a similar ratio of the computer and server assets. The text "The water/wastewater portion of the" is implied at the beginning of each of the cyber assets.

Assets that may be impacted by a malicious actor include:

Financial system. The software, hardware, and people to perform the financial processes that run payroll, generate invoices, receive payments (accounts receivable), receive invoices from vendors and approve payments (accounts payable), reconcile all transactions (general ledger), track inventory, process meter readings to issue invoices, and many other financial functions. The financial system may cover only the water/wastewater utility or the entire organization that may provide additional services, e.g., a municipality that provides police, fire, libraries, and water or wastewater utilities.

The financial health of the organization is not relevant to this asset. For example, the organization credit score or bond ratings are not relevant to this discussion.

Business enterprise systems/Information technology (IT) network. The combination of all computer hardware, software, and people to support the business enterprise. This includes all the servers, network equipment, and software that support computer systems for the normal business of the enterprise. Typically, the IT networks hold these and many other systems, e.g., email, access to the Internet, printing, file sharing, authentication (usernames/passwords), Geographic Information Systems (GIS), Computerized Maintenance Management Systems, consumption meter reading systems, customer (or citizen) information systems, water quality sampling results, spreadsheets, documents, and reporting systems.

Some assessors may want to divide the IT infrastructure into the hardware and the software components as individual assets if they are identified as single points of failure or if they have a drastically different profile than the other portions of the system.

Business enterprise applications (software). These are all the systems that support email, access to the Internet, printing, file sharing, and the normal business of the enterprise that is completed with computers. Typically, GIS, meter reading systems, work order management, water quality sampling results, reporting systems, and many other software packages are found on the IT networks.

Business enterprise infrastructure (hardware). The hardware can be physical, virtual, or cloud based. Domain controllers, file servers, network switches, point-to-point radios, network firewalls, email servers, phone systems, and other corporate wide information technology support infrastructure. The infrastructure could be physically on-premises or hosted in a cloud environment.

Process control system (PCS). These systems may be called operational technology (OT), industrial control systems (ICS), supervisory control and data acquisition (SCADA), distributed control systems (DCS), or Industrial Internet of things (IIoT). For this discussion, these terms are interchangeable, and the term PCS is used throughout the document. The PCS is the people, hardware, and software including the human machine interface (HMI), historian databases, sensors, remote terminal units, programmable logic controllers (PLC), distributed control units, network infrastructure, and data radios that allow for the automation of the water and wastewater processes. This includes the monitoring of sensors and controlling of equipment such as pumps, valves, and other water or wastewater processing equipment at a treatment facility, in the water distribution system, or wastewater collection system. PCS devices that are connected through cellular networks to provide information are in this category.

Consequences

The consequences in J100-21 are made up of four categories:

- fatalities,
- injuries,
- utility financial consequences, and
- regional economic consequences.

It is assumed for all the IT cyberattacks that there are no fatalities, injuries, or regional economic consequences. There are only utility financial consequences. PCS cyberattacks may need to take fatalities, injuries, and regional economic consequences into consideration.

The organization should use the same consequence estimate methodologies as physical assault or natural disasters. The consequence categories typically include:

- initial recovery—return of the most critical functions in a temporary location.
- return to normal—repair the asset and move from temporary location.
- make improvements—rebuild the asset to better specifications.

For cyber, there are a few standards that have consequence categories adjusted from the general categories shown previously. Two are the National Institute of Standards and Technology (NIST) Special Publication (SP) 800-61 and NIST Cybersecurity Framework (CSF).

NIST SP 800-61 Computer Security Incident Handling Guide includes the following:

- containment, eradication, and recovery
- post-incident activity.

NIST CSF includes the following:

- detect
- respond
- recover.

Consequence estimates include the effort hours and software costs for containment, eradication, response, and recovery. Recovery is defined for cyber to be a return of pre-incident functionality.

Post-incident activity from NIST SP 800-61 can be anything as small as a post-incident review meeting to the replacement of every piece of computer hardware and building a cybersecurity program. Post-incident activities and improvements to the assets or their protection systems are not included in the consequence estimates presented here.

All dollar amounts are based on 2021 values unless otherwise indicated. Consider adjusting for inflation as needed.

Organizations that have cyber insurance should consult with their cyber insurance company on the level of coverage for each of the threat–asset pairs. Consider the deductible and the organizational effort to recover the services that are not covered by insurance.

There is not a distinction between an insider threat and outsider threat in this document beyond this paragraph. The evaluator may choose to include cyber insider and cyber outsider as separate versions of each of the threats. The consequence of the insider may be slightly higher compared to an outsider. The vulnerability to an insider may be much higher. The threat likelihood of the outsider may be much higher.

For each threat–asset pair, consider if the organization will activate a virtual, local, or regional emergency operation center (EOC). This activation will consume effort hours for upper management, emergency management, and public information officials. Consider adding the cost of ramping up the EOC to the manually calculated consequences presented in this document.

SECTION B.2:   DEFINITIONS

1. *Controls:* Cybersecurity topics or questions about cybersecurity topics.

2. *"Cost of a Cyber Incident: Systematic Review and Cross-Validation" (CCI):* A document created by cisa.gov in October 2020.

3. *Critical Infrastructure Security Agency (CISA):* An agency in the DHS of the US Federal Government (cisa.gov). Works with partners to defend against threats and collaborates to build a more secure and resilient infrastructure in the future.

4. *Cyber Security Framework (CSF):* A cyber assessment method from NIST.

5. *Demilitarized zone (DMZ):* In computer security, a demilitarized zone or perimeter network is a network area (a subnetwork) that sits between an internal network and an external network. The purpose of a DMZ is that connections from the internal and the external networks to the DMZ are permitted, whereas connections from the internal network to the external network are blocked. The DMZ has host computers with connections to the external network and the internal network. This allows the DMZ's hosts to provide services to the external network and internal network while protecting the internal network. For someone on the external network who wants to illegally connect to the internal network, the DMZ will stop that connection. Please see the Purdue network model for more information.

A process control system (PCS) DMZ is used to separate the IT network from the PCS network. Information in the PCS that is needed by users on the IT network is provided in the PCS DMZ.

6. *Department of Homeland Security (DHS):* A department of the US federal government. Safeguards the American people, homeland, and values.

7. *Distributed Control System (DCS):* The entire system from sensors to historian that allows automated operations. Referred to as PCS in this operational guide.

8. *Emergency operation center (EOC):* The physical or virtual location at which the coordination of information and resources to

support incident management (on-scene operations) activities normally takes place.

9. *Industrial control system (ICS).* The entire system from sensors to historian that allows automated operations (referred to as PCS in this operational guide).

10. *Information technology (IT):* The combination of all hardware, software, and people to support the business enterprise computer systems.

11. *Laboratory Information Management System (LIMS):* Software to capture hand-grab water quality sample results and generate reports.

12. *National Institute of Standards and Technology (NIST):* A division of the US Commerce Department.

13. *Operational technology (OT):* The entire system from sensors to historian that allows automated operations (referred to as PCS in this operational guide).

14. *Operator interface terminal (OIT):* A device that connects to a PLC to operate the local equipment or system. It is typically panel mounted.

15. *Process control system (PCS):* The entire system from sensors to historian that allows automated operations.

16. *Programmable logic controller (PLC):* This device takes inputs from sensors, makes decisions, and issues commands to control elements such as pumps.

17. *Public information officer (PIO):* The PIO is the individual responsible for communicating with the public and media and/or coordinating with other agencies, as necessary, with incident-related information requirements.

18. *Special Publication (SP):* Documents created by NIST.

19. *Supervisory control and data acquisition (SCADA):* The entire system from sensors to historian that allows automated operations (referred to as PCS in this operational guide).

20. *Water/Wastewater Agency Response Network (WARN):* A network of utilities helping other utilities to respond to and recover from emergencies.
https://www.awwa.org/Resources-Tools/Resource-Topics/Water-Wastewater-Agency-Response-Network

21. *Water Health and Economic Analysis Tool (WHEAT):* Designed to assist utility owners and operators in quantifying an event's public health consequences (i.e., injuries and fatalities), utility-level financial consequences, direct and indirect regional economic consequences, and downstream impacts.

SECTION B.3:   THREATS AND CONSEQUENCES

The following sections present cybersecurity threats and the associated consequences. Each threat is presented with a definition and consequence.

Cyber Theft or Diversion

Definition: Cyber theft or diversion. A malicious actor has communicated with the utility and has convinced an employee to redirect an electronic funds transfer, automated clearing house payment, or paper check to the malicious actor. Redirection of an individual employees' single payday could be covered in this threat. However, an individual's single payday is generally considered to be a smaller consequence compared to the redirection of a single payment to a vendor.

Physically intercepting a paper check and changing the payee to the malicious actor's name or company (check washing) is part of this threat.

Consequence: Cyber theft or diversion of the financial system.   The consequence is defined as the worst reasonable scenario, typically the highest regularly scheduled reoccurring payment. First consideration should be given to a construction project progress payment. These may be scheduled monthly for a year or more, and the construction contractor may be a new vendor for the organization, so the identity of the malicious actor asking for the change may be more easily falsified. Other reoccurring payments to consider are:

- raw water payment
- treated water payment
- electricity utility payment

- chemical payment
- other vendor/contractor payment.

The consequence is the dollar amount of the highest regularly scheduled payment and any costs associated with recovering the payment as well as its economic impacts.

Cyberattack of Business Enterprise/IT Network

Definition: Cyberattack of business enterprise/IT network. This is assumed to be a ransomware attack where a malicious actor sends an email with a malevolent payload, or a malicious file is installed onto the network (potentially via USB flash drive, a downloaded malicious file update, or malicious file from a DVD/CD). This attack will encrypt the servers and computers with a demand for payment.

Many IT organizations keep "operational backups" and "offline backups." Operational backups are kept on a live server and are used to restore accidentally deleted files. Offline backups are moved to removable media or powered-down file shares that are moved to other remote secure locations. It is assumed that the operational backups are encrypted by the ransomware attack, and the offline backups are not encrypted by the ransomware attack.

It is assumed that data exfiltration has occurred, and the malicious actor has a copy of all data from the financial system or other sensitive data.

Stand-alone extraction of personally identifiable information (PII) such as Social Security numbers, bank account information, and credit card information without encryption may be captured as an additional threat in this category. Definitions of PII vary by state or county. Please consult the local attorney general for PII definition and reporting requirements.

Consequence: Cyberattack of business enterprise/IT network. The consequence of a ransomware attack on the IT network is based on whether the ransom is paid and the effort of the response. The response generally includes:

- Save a clone (produce a bit-by-bit duplicate) of critical assets as evidence for law enforcement by disconnecting the Ethernet cables and keeping the computer or server powered on.

- Notification of law enforcement.
- Pull spare hardware from inventory.
- Factory reset hardware that will be reused.
- Rebuild systems/equipment from known good backups in a test network and watch for reinfection.

It is assumed that the network equipment does not need to be factory reset and restored from offline configuration backups, though this is a good idea to do. The value of the loss of any intellectual property is not included in the estimates in this document but should be included in the assessment.

The following methods are presented for the evaluator to choose from. Method 1 is strongly recommended. If Method 1 is not used, document the reason why, and the justification for selecting a different method or approach to calculate the consequence.

Method 1: Add up the costs. Calculate the effort to detect, isolate, and factory reset and restore every server and every computer from backup with the available resources. Restoration will likely be prioritized for different services such as police, fire, payroll, reading meters, processing meter invoices, issuing purchase orders, library, parks, and many others. Prioritization of recovery order may already be covered in the organization's Continuity of Operations Plan (COOP).

Discuss with the IT network support and or the server support group the effort to restore a server, computer, and an endpoint device. If no values are available, assume that it takes eight (8) hours to reset and restore a server and two (2) hours to reset and build a computer.

Discuss with the application owners the time needed to find then hand enter the data that may be lost since the last offline back up was created.

Add the costs of these actions that may occur:

Containment and Eradication

- Notify upper management and PIO.
- Notify law enforcement and support them if they arrive.
- Activate and staff an EOC.
- Utilize the AWWA WARN system or third party for additional staffing.

- Disconnect the Ethernet cables from one physical server and leave it powered on for law enforcement.
- Stop the cyberattack. Conduct threat hunting, find the entry point, and block further intrusions.
- Disconnect the Ethernet cords or power down all impacted servers, computers, and equipment.

Initial Recovery

- Factory reset of servers, computers, and endpoint devices to be reused.
- Consider factory reset of and restoration of network switches and routers.
- Restore from backup to equipment that is on a disconnected test bench and scan for viruses. Watch for reinfection.
- Move the equipment to the production IT network.
- Continue to monitor for delayed infections.

Return to Normal

- Hand entering data that was not able to be restored.
- Restore redundancy.
- Provide two years of credit monitoring for individuals in the billing system with PII.
- Consider recommending all users change their passwords. Change system and network passwords.

Make Improvements

- Update policies and procedures.
- Conduct training.
- Review the security technology stack and consider third-party vendor support.
- Perform cybersecurity risk and resilience assessment.

Service Interruption Impact

- Lost revenue due to service denial.
- Cost of paying the ransom. Estimate the ransom to be the insurance coverage limit provided by the organization cybersecurity insurance policy.
- Liabilities for service interruption.
- Contract defaults.

Table B.1 Cost of cyber incident based on budget of the utility

Factor	Minimum	Average All Sectors	Average Public Sector	Maximum
Cost-to-budget ratio	0.25%	0.37%	0.80%	1.00%
Dollar	$100,000	$5,100,000	$5,900,000	$30,000,000

If Method 1 is too cumbersome to calculate, consideration could be given to the following two methods presented in the order of accuracy. Document the reason a different method was selected.

Method 2: Ratio of cost of cyber incident (CCI) to revenue or budget. The Cybersecurity and Infrastructure Security Agency (CISA) is an agency under the Department of Homeland Security (DHS). CISA has studied the cost of cyber incidents and has published the Cost of Cyber Incidents. This document presents "the cost-to-revenue ratio… for the public sector is 0.8% for the 75th percentile, and approximately 2% for the 85th percentile. The median cost-to-revenue ratio across all of the sectors is 0.37%."* Please check https://www.cisa.gov for information updates. This document says, "The accurate estimation of losses and risk quantification is not a trivial issue in cybersecurity."

Public sector includes federal agencies, state universities, county governments, and municipal governments. Whether the water/wastewater utility is private or public, the "Average Public Sector" value of 0.80% is recommended.

These numbers tend to be more accurate for small-(<250 employees) and medium-sized (<1,000 employees) organizations. Larger organizations should use Method 1.

Method 3: Ratio between the total cost of the breach and the number of records affected. The CCI states, "The weak association between the (total cost of breach and number of people/records affected) implies that relying on per-record estimates to approximate total breach costs is not appropriate…" The CCI shows there is research that the cost per record for a breach fluctuates between $0.58 per record and $201.00 per record. This wide range is not useful to estimate a consequence. Check to see if there is more recent information.

---

* https://www.cisa.gov/sites/default/files/publications/CISA-OCE_Cost_of_Cyber_Incidents_Study-FINAL_508.pdf

Table B.2 IT ransomware impact based on utility population served

Population	No – Ransom Paid	Yes – Ransom Paid
Greater than 100,000	$600,000	$1,200,000
50,000–99,999	$400,000	$800,000
3,300–49,999	$200,000	$400,000
Less than 3,300	$100,000	$200,000

This method is based on the number of people that are impacted. The America's Water Infrastructure Act of 2018 (AWIA) divides the water providers into four categories based on population served. Select the appropriate value from the table. There is no external reference for this information. It is anecdotal experience from the J100-21 cybersecurity subcommittee.

Table B.2 doubles the consequence if the ransom is paid based on the "Cost of a Data Breach Report 2021" created by IBM (https://www.ibm.com/security/data-breach). (Please check the website for any report updates.)

Cost of loss of personally identifiable information/customer financial data. A consequence of $180 per record is recommended by the IBM report. This includes the following services:

- administration time
- attorneys – internal and external
- notification of breach
- credit monitoring.

Assume PII data was copied from the billing system and became publicly available or was made available for sale. The consequence is the effort to notify all potentially impacted people and provide them with two years of credit monitoring at the utility's expense. Assume several hours to notify state law enforcement and file a report. Assume 5 min to notify each impacted person using an automated system such as email. Assume $60 per person for a year of credit monitoring in 2021.

Consequence = (Count of records × $180 per record) + (Number of people × $60/year × 2 years)

Cost of loss of patented information or trade secrets. Consult with legal and senior management to estimate the financial value of the patented information or trade secrets that were copied from the system.

75

Summary for Consequence: Cyberattack on the Business Enterprise/IT Network

Method 1 should be selected for a cyberattack on the IT network. If Method 2 or Method 3 is selected, document the reasons why Method 1 could not be used.

Sum all the costs of the consequences:

- cost of the ransom if it was paid
- cost of ransomware removal
- cost of loss of PII
- cost of trade secrets
- cost of credit monitoring

Cyberattack on the PCS Network—Ransomware

Definition: Cyberattack on the PCS network—ransomware. This is assumed to be a ransomware attack where a malicious actor sends an email with a malevolent payload and pivots to the PCS DMZ then pivots to the PCS network, or a malicious payload is carried onto the PCS network with a USB thumb drive. This attack will encrypt the PCS servers and computers with a demand for payment.

Consequence: Cyberattack on the PCS network—ransomware. Ransomware released on the PCS network will encrypt the HMI screens and related servers including the PCS historian. It is assumed that the programmable logic controllers (PLC) programming and network switches configurations are not impacted. There are no known published guidelines for estimating the impact of ransomware on the PCS network. Manual calculation to add up the costs is recommended.

Method 1: Add up the costs. Estimate the number of hours the PCS support team needs for containment, eradication, and initial recovery of PCS functionality. The PCS support team will need to provide a time frame to perform these actions. There will likely be additional "on-call" costs for both PCS support team and operations staff while proper operation of the system is confirmed.

If no estimation is available from the PCS support team, antidotal experience from the authors of this document suggests considering 75 hours total for an HMI Server, a historian server, and ten SCADA

76

clients. The automated systems will need to be run in manual for the duration of the outage.

If a utility does not have an estimate for the number of licensed operators needed to run in manual, consider using the count of on-watch operator(s) and multiply by three individuals for the duration of the outage provided. Overtime pay rates should be used.

The following is a detailed list of actions for consideration:

Containment and Eradication

- Notify the upper management and PIO.
- Notify law enforcement and support them if they arrive.
- Activate and staff an emergency operations center.
- Utilize the AWWA WARN system or third party for additional staffing.
- Consider contacting the PCS Cybersecurity Emergency Response Team (CERT) for assistance.
- Place automated equipment in manual.
  - Additional qualified operators may need to be placed at strategic locations at treatment facilities.
  - A distribution system or collection system may need additional individuals to be stationed at tanks or pump stations to monitor tank levels, pressure, and operations.
  - Roving watches may need to be increased to visit sites more frequently.
  - Extra sampling and paper documentation may be required to meet permit requirements if the PCS system is not recording online water quality sensor values.
- Disconnect the Ethernet cables from one physical server and leave it powered on for law enforcement.
- Stop the cyberattack. Start threat hunting, finding the entry point, and blocking further intrusion.
- Power down impacted computers and equipment.

Initial Recovery

- Factory reset of servers, computers, and end point devices to be reused.

- Consider factory reset of and restore network switches and routers.
- Consider factory reset and restore PLCs, Operator Interface Terminals (OITs), and Ethernet-based sensors.
- Restore from backup to equipment that is on a disconnected test bench and scan for malware. Watch for reinfection.
- Move the equipment to the production PCS network.
- Perform point-to-point checks to verify field values are accurate at the local displays, HMI screens, and historian.
- Move systems from manual to automatic operation and verify operations occur as expected.

Return to Normal
- Restore redundancy.
- Consider recommending all users change their passwords.

Make Improvements
- Update policies and procedures.
- Conduct training.
- Review the security technology stack and consider third-party vendor support.
- Perform cybersecurity risk and resilience assessment.

Other Impacts
- Costs of any water that could not be sold.
- Penalties for any contract violations with customers or consumers.
- The ransom if it is paid.

Add the cost of the PCS support team and the cost of the additional staff effort at overtime rates to estimate this consequence.

Method 2: Consequence estimates. If Method 1 is too cumbersome to calculate, Table B.3 indicates the anecdotal experiences of the J100-21 cyber team. The doubling of the consequence if the ransom is paid is based on the IBM "Cost of a Data Breach Report 2021."

Choose the appropriate value from the table for PCS system consequence. There is no external reference for this information. It is anecdotal experience from the J100-21 cybersecurity subcommittee.

Add the expense of the ransom if it is paid.

Table B.3 PCS ransomware impact based on utility population served

Population	No – Ransom Paid	Yes – Ransom Paid
100,000 or greater	$300,000	$600,000
50,000–99,999	$200,000	$400,000
3,300–49,999	$100,000	$200,000
Less than 3,300	$50,000	$100,000

Summary Consequence: Cyberattack on the PCS Network

Method 1 should be selected for a cyberattack on the PCS network. If Method 2 is selected, document the reasons why Method 1 could not be used.

Cyberattack on the PCS Equipment—Sabotage Operator Changes

Definition: Cyberattack on the PCS equipment—sabotage operator changes. A malicious actor gains remote access to the PCS network and can manipulate the system with the same access level as an operator. It is assumed that mechanical safety features and PLC safety logic programming work as intended to protect the system. Alarming will function as programmed; however, the malicious actor will likely be able to acknowledge alarms as they occur.

Discuss with the senior operators what reasonably could be done to injure the population as the logged-on operator. Their experience and knowledge of their system will be more valuable than the options presented here.

Worst reasonable cases to consider may include:

- Changing chemical levels—Manipulate chemical feed rates to a minimum or to a maximum feed rate.
- Depressurization of the distribution system—Stopping pumps for an extended amount of time leading to a lowering of pressure in the distribution system below acceptable levels.

Consequence: Cyberattack on the PCS equipment—sabotage operator changes. The experience of the utility experts is valuable and takes priority over any estimates here.

The US Environmental Protection Agency (USEPA) Water Health and Economic Analysis Tool (WHEAT) calculator may be used to estimate a consequence. WHEAT provides fatalities, injuries, utility consequences, and regional economic consequences. Data the organization has for regional economic consequences will be more accurate than the WHEAT calculator. Examples of regional economic consequences to be considered:

- dialysis centers that may have clients injured
- structures that burn down while there is limited water for firefighting
- electrical generation facilities that may lose cooling water replenishment
- contracts with local bottling companies or other large water users
- bad publicity leading to lower tourism
- losses to agriculture.

Generally, the corporate or private individuals and their insurance companies are liable for these losses and may be excluded from the consequence estimates in this document.

Changing chemical levels. Consider these factors to estimate the consequence:

- What is the maximum or minimum value that an operator could change the chemical feed rates?
- How long before automated action would occur to isolate water with dangerous chemical levels?
- What is the response time for the PCS systems to alarm?
- What is the response time for the plant operator after the notification is received?
- Would individuals potentially get poisoned or injured with dangerous chemical levels?
- How many hours to restore normal chemical addition?
- What needs to be done with the water that is out of acceptable levels?
- What notification is needed for hospitals and other medical facilities?
- What notification is needed for the public and local regulatory agency? Boil water? Do not use?
- Does bottled water need to be brought in?
- Under what conditions would state or federal assistance (FEMA) be requested?

Depressurization of the distribution system. The damage of depressurization is mitigated by venting/vacuum break valves (VENT/ VAC). If the organization does not have documented maintenance records of the VENT/VACs, there is the possibility that a vacuum would be drawn, and the pipes would collapse in the vacuum as the water is drained through normal usage.

Even with successfully functioning VENT/VACs, restarting the pumps and filling sections of the distribution system will need to be done with care. Coliform detection testing (Bac-Ts) may need to be performed. Public notification and water use restrictions may be needed.

The experience of the senior water operators and engineering staff is needed to estimate these consequences. Consider the following for calculating the cost of depressurization:

- The cost of providing bottled water, public showers, or toilet facilities to neighborhoods and other users.
- The cost of public notification and internal coordination.
- The cost of replacing damaged piping, sanitizing, and placing back online.
- Are there liabilities for water not being available for firefighting and structures burning down?
- Are there liabilities for patients in hospitals and kidney dialysis centers that may need to be relocated?
- Are there liabilities for potential closing of industrial and commercial companies that need water?
- Under what conditions would state or federal assistance (FEMA) be requested?

WHEAT Calculator. This consequence may be very challenging to estimate. The WHEAT calculator needs the following information to create an estimated consequence for a damaged distribution pipe:

- the length of pipe affected (in feet)
- the diameter of the pipe
- the cost of the asset
- the duration of the service outage (days)
- the percent of the population without service.

81

No fatalities or injuries are expected. Attempt to estimate the loss of credibility, public notification, and continued follow-up.

Summary Consequence: Cyberattack on the PCS Equipment—Sabotage Operator Changes

Add all the costs together to determine the consequence.

Cyberattack on the PCS Equipment—Sabotage Programming Changes

Definition: Cyberattack on the PCS equipment—sabotage programming changes. It is assumed that it is not possible for anyone to release hazardous chemicals to the environment from the PCS network even with full knowledge of the chemical system and full administrator access to the HMI programming software and related PLCs and PLC programming software. If it is possible to release hazardous amounts of chemicals to the environment from the PCS network, select this as the reasonable worst case. Consider using EPA WHEAT to estimate a consequence based on the chemical, the amount of that chemical that can be released, and the impacted population.

This scenario could be considered a "worst case" for PCS sabotage. J100-21 states "worst reasonable case." Some individuals may consider the following scenario too "unreasonable" for evaluation. Consider using the concepts presented and create a reasonable scenario for your utility. Remember, this exercise is to estimate the consequence in dollars with the exiting protections. How often this is attempted is covered in the Cybersecurity Threat Likelihood section (B.5) of this appendix.

A malicious actor can remotely change the PLC programming. The malicious actor may be able to modify or override the safety logic in the PLC. They would be able to change cyber-physical safety features. For example, they may be able to start all pumps that may lead to an overpressure situation that results in main breaks. However, they would not be able to impact the operation of electro-mechanical protections. For example, the malicious actor would not be able to change the mechanical pressure relief valve(s) setpoints or stop them from functioning. Consideration should be given to the possibility that the malicious actor may have access to multiple PLCs and could place timers in multiple PLCs for actions to happen simultaneously at multiple locations. For example, all pumps in all locations in the distribution system could be started at the same time, leading to over pressurization and multiple main breaks.

Assume that once the malicious actor has finished changing the logic programming and setting delay timers, they may change the logic controller passwords—mandating the factory reset and restore from offline backups of the logic controller—then release ransomware on the PCS network that would encrypt or corrupt the HMI servers and clients. This may delay the detection of the damaged distribution system and delay the response time to factory reset the logic controllers and restore the programs from offline backups.

If this scenario seems unreasonable, coordinate a discussion with the individuals with the most knowledge about operating the water/wastewater systems, the engineering of the system, and the PCS network. Ask them to brainstorm what is a "reasonable worst case" that someone with their knowledge could come up with to cause damage to the population with only access to the computers. Assume the utility supervisors, engineers, and PCS system experts are available to respond and recover while the scenario is developing. An organization may perform a Consequence-driven Cyber-informed Engineering (CCE) assessment to determine the reasonable worst cases, also referred to as High Consequence Events. Information can be found at the Idaho National Laboratory website. (The link in 2021 is https://inl.gov/cce/).

Programming changes to the PCS firewalls and the PCS network switches are covered under this threat. Generally, changes to these have a smaller consequence than changing PLCs.

Consequence: Cyberattack on the PCS equipment–sabotage programming changes. If a reasonable worst case was not created, consider selecting a "reasonable worst case" scenario for your organization from these options:

- overpressurization of the distribution system
- turn off wastewater treatment plant(s)
- turn off sewage lift stations.

Common to all these threats is the time to factory reset the PLCs and restore the program from an offline backup. The cyberattack on the PCS network (ransomware) consequence would be added to this consequence. Local manual operation of the equipment and testing the return of the PLCs will take a significant amount of time.

Consider using the USEPA WHEAT calculator to estimate a consequence. WHEAT provides utility consequences and regional economic consequences. The utility consequences are used for the consequence. The regional economic consequences are generally not included in the consequence calculation. The utility experience will be more accurate than the regional economic consequences provided in WHEAT. Examples of regional economic consequences to be considered could be:

- dialysis centers that may have clients injured
- structures that burn down while there is limited water for firefighting
- electrical generation facilities that may lose cooling water replenishment
- contracts with local bottling companies or other large water users
- bad publicity leading to lower tourism
- losses to agriculture.

Generally, the corporate or private individuals and their insurance companies are liable for these losses and may be excluded from the consequence estimates.

Overpressurization of the distribution system. Each distribution system is different, and there is no known guidance for the number of main breaks during an overpressure condition. The experience of the most senior operators and engineers in your organization is valuable in this estimation and would take priority over these estimates.

It may be reasonable to assume that 10 main breaks would happen within a one-mile radius of the discharge of the pumps. Five main breaks may occur in the next mile of mains for a total of 15 main breaks at each pump station location.

Assuming that piping and excavation equipment is readily available, it is estimated that it would take 12 h to fix each main break. Coliform detection testing (Bac-T) would be needed to put sections of the distribution system back into service. The distribution system would come back in segments over time. For ease of calculation, it is safe to assume that the distribution system would come back online at the end of the time needed to repair all the breaks. Public notification and boil water alerts may be needed.

Consideration should be given to:

- the cost of repairs
- the water service outage duration
- the percent of the population that is without water for that duration
- the $/1,000 gallons of water not sold
- the economic impact to the community due to loss of water for firefighting, wastewater operations, etc.

The WHEAT calculator may be useful to estimate this consequence.

Turn off wastewater treatment plant(s). It may be possible to turn off the treatment plant. Many wastewater plants are end-of-line facilities, and sewage overflow is a concern. Raw sewage can be life-threatening if humans are able to interact with it. In addition, raw sewage being discharged to a beneficial use waterway could impact the local communities in numerous ways.

To estimate this consequence, consider:

- PCS support team effort to factory reset the PLCs and SCADA servers and restore from backup
- run facilities in manual
- activate the EOC and public notifications
- effort to handle the overflowing sewage
- report to environmental agencies
- pay fines that may be assessed.

Turn off sewage lift stations. It may be possible for a malevolent actor to turn off an individual or all sewage lift stations from the PCS. This may cause wastewater to overflow into public locations. If available, use the Capacity Management Operation and Maintenance (CMOM) report to determine the order that sewer entry holes may overflow. Consider where sewage would go from the offline lift station(s).

To estimate this consequence, consider:

- PCS effort to factory reset the PLCs and restore from backup
- run the facilities in manual
- effort to handle the overflowing sewage
- report to environmental agencies
- pay fines that may be assessed.

Add up the costs. Add all the costs together to determine the consequence.

SECTION B.4:   CYBERSECURITY VULNERABILITY

Introduction

Vulnerability is the effectiveness of countermeasures to protect an asset from a defined threat. Therefore, it is an inherent state of a system (e.g., physical, technical, organizational, cultural) that can be exploited by an adversary or impacted by a natural or dependency/ proximity hazard to cause harm or damage. A vulnerability estimate is determined for each threat–asset pair. J100-21 vulnerability has a range of 0.00 to 1.00.

The vulnerability estimate is interpreted as the conditional probability that a specific threat will cause the specific estimated consequences to the asset. For malevolent threats, it may be interpreted as the probability of a malicious actor's success, given the threat occurs.

A minimum vulnerability of 0.03 was selected for cyber threat– asset pairs as the best possible score because no cybersecurity system is perfect, and malicious actor capabilities are always improving.

50 to 1,000 Controls to Evaluate

Cybersecurity vulnerability is determined with an assessment that covers many aspects of cybersecurity that are called controls. There are several cybersecurity standards that have anywhere between 50 and 1,000 controls. The implementation of each control is assessed, and a score is given to each control.

There are control questionnaires that are focused on information technologies. Examples include:

- NIST SP 800-53 Security and Privacy Controls for Information Systems and Organizations
- NIST Cybersecurity Framework (CSF)
- ISACA Control Objectives for Information Technologies (COBIT)
- International Organization for Standardization/International Electrotechnical Commission 27000 Information Security Management Systems family of controls.
- SANS Institute
- Center for Internet Security Critical Security Controls

There are control questionnaires that are focused on process control systems. Examples include:

- NIST SP 800-82 Guide to Industrial Control Systems
- International Society of Automation (ISA)/International Electrotechnical Commission (IEC) 62443 Security for Industrial Automation and Control Systems family of controls
- Advanced Cyber Industrial Control System Tactics, Techniques, and Procedures (ACI TTP) for Department of Defense (DoD) Industrial Control Systems (ICS)

There are questionnaires that combine IT and PCS networks:

- American Water Works Association Water Sector Cybersecurity Risk Management Guidance and Self-Assessment Tool[†] (AWWA Tool)

All these standards have versions. The version numbers are intentionally withheld in this document. Please refer to the appropriate website for the most recent version.

Some of these standards have the capability to select a "security assurance level" that will drive the number of controls or questions in each of the topics discussed.

If your organization has used a national or global cybersecurity standard before, the organization may want to consider using the same standard to measure the progress since the last assessment.

It is acceptable for an assessment team to select different standards for different groups of asset–threat pairs. For example, if an organization

---

[†] https://www.awwa.org/Resources-Tools/Resource-Topics/Risk-Resilience/Cybersecurity-Guidance

has used NIST CSF for a previous enterprise assessment, that standard may be used for the enterprise-related asset–threat pairs and NIST 800-82 could be used for the PCS-related asset–threat pairs.

The standards listed previously may or may not provide an overall score that can be converted to a vulnerability estimate for use in the J100-21 methodology. However, the Department of Homeland Security (DHS) created the Cyber Security Evaluation Tool (CSET) that will provide an overall score for any of the previously listed standards and many other standards.

CSET

CSET was originally released in 2009 and is actively supported and updated regularly by CISA. CSET is a computer program that is run locally on a computer. The answers do not need to leave the computer that it is running on. Evaluators may select several parameters that will determine the number of controls or questions asked including a security assurance level and the ability to select multiple standards that can be assessed simultaneously.

CSET provides an "overall score." The caveat is that the overall score from CSET has 0% as the worst possible score and 100% as the best possible score. Thus taking 100% minus the CSET score will provide the J100-21 vulnerability score. The 3% minimum for cyber vulnerability does not need to be considered unless the CSET provides a very high score.

AWWA and DHS have worked to integrate the AWWA Tool with CSET. The evaluator needs to complete the AWWA Tool assessment(s) online. This creates a spreadsheet as output. The AWWA Tool spreadsheet results may be imported into CSET. CSET generates an overall score along with several reports.

AWWA Water Sector Cybersecurity Risk Management Self-Assessment Tool

None of the presented standards for cyber assessments is focused solely on water/wastewater. AWWA saw the need to focus on cybersecurity in water/wastewater and formed a team to create an assessment using the language of controls, questions, and examples for water and wastewater utilities into a single questionnaire. The AWWA Tool generates a spreadsheet that is customized to the utility's cyber profile and available technologies; it has approximately 100 assessment questions. The AWWA tool is available free of charge to individuals that provide an email address to AWWA.

The AWWA tool combines the IT and PCS questions, which may be appropriate for organizations that only have one network and a common IT and PCS support group and only need one vulnerability score. For organizations that need multiple vulnerability scores, it is recommended to run the AWWA tool for each threat–asset pair.

The assessment group may want to consider grouping similar threat–asset pairs to generate a vulnerability score that will be used for the group. For example:

- The AWWA tool could be run for the Business Enterprise System and IT support group responsible for the billing system under a theft/diversion attack.
- The AWWA tool could be run again for the PCS and its group under a ransomware attack.

While doing each threat–asset pair individually would take some time (although it may be a very fast learning curve with experience), the total number of attacks and the total number of assets can be constrained. As noted, three assets under as few as three types of threats might work, e.g.:

- Theft/Diversion from financial system
- Ransomware on IT network
- Ransomware on the PCS network
- Operator access sabotage of PCS network

The AWWA tool does not provide a vulnerability score in the tool itself or in the output of the spreadsheet as of 2021. This may change in the future.

In the meantime, the next section describes how to utilize the AWWA Tool in a completely manual process to calculate a Vulnerability score with the added benefit that values for Detect, Delay, and Respond are determined.

Manually Calculating Vulnerability with the AWWA Tool

There are approximately 100 control measures in the AWWA Tool. The online version of the AWWA tool has introductory questions about technology and will eliminate most of the Not Applicable controls. However, this section documents how the assessment is done manually, and it is incumbent upon the user to identify controls that are not applicable to their system. An answer value of "0 – Not Applicable" is allowed when manually calculating vulnerability. The manual vulnerability assessment worksheet is in Sec. B.7.

Priority Options

There are weights applied to the priority values and the status values of the AWWA Tool assessment. The weights presented can be adjusted by the assessment team.

Based on cross-sector best practices, each control is assigned a priority in the AWWA Tool, one through four, with one being the highest priority (e.g., the most protective). Based on this priority, a weighting is applied. Two options are presented here. Option 1 places more weight on the Priority 1 questions. If the basics are covered, the assessment will score higher. Option 2 lowers the weight of the Priority 1 questions and raises the weight of the other lower-priority questions. The assessors can use any weights desired. Option 1 is used in the examples and appendixes. The presented options are summarized in Table B.4.

Implementation Status Options

The implementation status of each of the control measures is assigned by the assessment team. There are four potential implementation statuses that may be selected. These statuses are applied a weight. Three different weights are presented. Option 1 is the most balanced and is used throughout the examples and appendixes. Option 2 is more lenient and provides more credit for "thinking about" or responding ad hoc. Option 3 meets the textbook definition of vulnerability to assess the existing conditions, and no credit is given for lowering future vulnerability.

Table B.5 summarizes the implementation statuses and the assigned weight for each.

Demonstration Scoring

The AWWA controls are divided into families including but not limited to the ones shown in Table B.6.

Table B.4 Vulnerability priority weight options

Priority Value	Priority Description	Priority Weight Option 1	Priority Weight Option 2
1	These controls represent the minimum level of acceptable security for SCADA/PCS. If not already in place, these controls should be implemented immediately.	60%	40%
2	These controls should be implemented second because they have the potential to provide a significant and immediate increase in the [cyber]security of the organization.	25%	30%
3	These controls provide additional security against cybersecurity attack of PCS systems and lay the foundation for implementation of a managed security system. These controls should be implemented as soon as budget allows.	10%	20%
4	These controls are more complex and provide protection for more sophisticated attacks (which are less common); they also provide for managed security systems.	5%	10%

Table B.5 Control status weight options

Status Value	Status Answer Text	Status Weight Option 1	Status Weight Option 2	Status Weight Option 3
1	Not planned and/or not implemented – risk accepted	5%	10%	0%
2	Planned and not implemented	10%	20%	0%
3	Partially implemented	25%	30%	50%
4	Fully implemented and maintained	60%	40%	100%

There are several controls in each of the families. Each control/question is identified with a number following the family abbreviation. For demonstration purposes, three of the AWWA controls were selected to demonstrate how to estimate vulnerability manually. The example in this appendix shows the possible weighted scores for each control measure, and how they apply to Detect, Delay, or Respond.

The AWWA Online CRAT (as of 2021) does not use leading zeros in the control numbering. Leading zeros were added here for clarity and easy sorting. The selected controls for this demonstration are IA-01,

Table B.6 Control family abbreviations

Abbreviation	Family	Abbreviation	Family
AT	Awareness and Training	MP	Media Protection
AU	Auditing	PE	Physical Environmental Security
CM	Configuration Management	PM	Program Management
DS	Data Security	PS	Personnel Security
IA	Identification and Authentication	RA	Risk Assessment
IR	Incident Response	SA	System Acquisition
MA	Maintenance	SC	System Communication

PE-02, and AU-05. Scoring for the demonstration uses Prioritization Weight Option 1 and Status Weight Option 1.

IA-01 Demonstration

IA-01 in the AWWA Tool states "Access control policies and procedures established including unique user ID for every user, appropriate passwords, privilege accounts, authentication, and management oversight."

AWWA has set a priority of 1 for this control. The weight for a priority 1 is 60% or 0.6.

The evaluator may consider these different status answers.

The evaluator gives an answer value of three (3) for this control/ question. This corresponds to a weight of 25% or 0.25.

Table B.20 shows this control is applicable to Delay but not applicable to Detect or Respond.

The Delay scores are calculated with this formula:

Individual Control Score = Priority Weight × Answer Weight

IA-01 Control Score = 0.6 × 0.25 = 0.15

IA-01 Control Score of 0.00 is used for Detect.

IA-01 Control Score of 0.15 is used for Delay.

IA-01 Control Score of 0.00 is used for Respond.

PE-02 Demonstration

PE-02 in the AWWA Tool states "Secure areas protected by entry controls and procedures to ensure that only authorized personnel have access."

Table B.7 IA-01 status evaluator considerations

Answer Value	Status Answer Text	Evaluator Consideration	Answer Weight
1	Not planned and/or not implemented – risk accepted	There is no policy. Using workgroup authentication or application security. Passwords have not been changed in years. No plans to change this.	5%
2	Planned and not implemented	There is no policy. Using workgroup authentication or application security. Passwords have not been changed in years. We are looking at writing a policy and considering a central authentication capable with our software in the next 12 months.	10%
3	Partially implemented	There is password guidance for length and strength but not a signed policy. The organization uses a flavor of Kerberos with individual accounts and password requirements for applications that can use Kerberos. There are some applications that are not capable of Kerberos and the application handles user authentication. The application owner is responsible for having users change passwords. Do not know what the application owner requirements are. Will add detail to the guidance in the next 6 months.	25%
4	Fully implemented and maintained	There is a policy written and reviewed annually that states the requirement to have individual user accounts, define password requirements, and define user and administrator levels and what the authentication software is. Access lists are reviewed annually.	60%

AWWA has set a priority of 1. The weight for a priority of 1 is 60% or 0.6.

The evaluator may consider these different status answers.

The evaluator gives a score of four (4) for this control/question.

Table B.20 shows this control is applicable to Detect and Delay but not Respond.

Individual Control Score = Priority Weight × Answer Weight

PE-02 Control Score = 0.6 × 0.6 = 0.36

PE-02 Control Score of 0.36 is used for Detect.

PE-02 Control Score of 0.36 is used for Delay.

PE-02 Control Score of 0.00 is used for Respond.

Table B.8 PE-02 status evaluator considerations

Answer Value	Status Answer Text	Evaluator Consideration	Answer Weight
1	Not planned and/or not implemented – risk accepted	There is a server and main network switch is in a closet that is not lockable. There are no plans to improve this.	5%
2	Planned and not implemented	There is a server and main network switch is in a closet that is not lockable. This room is included in a physical security upgrade project that will occur in the next 12 months.	10%
3	Partially implemented	There is a data center with a card-read door. The access list is reviewed annually. Some network equipment is in closets that are not lockable, but the perimeter of the building is always locked.	25%
4	Fully implemented and maintained	There is a data center with a card-read door. The access list is reviewed annually. The network closets have steel keys, and the IT department has documented the individuals with the keys.	60%

AU-05 Demonstration

AU-05 in the AWWA tool states "Risk-based business continuity framework established under the auspices of the executive team to maintain continuity of operations and consistency of policies and plans throughout the organization. Another purpose of the framework is to ensure consistency across plans in terms of priorities, contact data, testing, and maintenance."

AWWA has set a priority of 2. The weight for a priority of 2 is 25% or 0.25.

The evaluator may consider these different status answers.

The evaluator gives a score of three (3) for this control/question.

The example shows this control is applicable to Respond but not Detect or Delay.

The respond scores are:

Score = Priority Weight × Answer Weight

AU-05 Control Score = 0.25 × 0.25 = 0.0625

AU-05 Control Score of 0.000 is used for Detect.

AU-05 Control Score of 0.000 is used for Delay.

AU-05 Control Score of 0.0625 is used for Respond.

Table B.9 AU-05 status evaluator considerations

Answer Value	Status Answer Text	Evaluator Consideration	Answer Weight
1	Not planned and/or not implemented – risk accepted	The supervisors know what to do or will figure it out in the moment. There is no plan to document this.	5%
2	Planned and not implemented	The IT department and PCS support department take regular backups. May occasionally restore an accidentally deleted file. No effort has been made to attempt to restore an entire server. But will in the next 12 months. The accounting teams and operators plan on discussing how to do things manually if the computers are not functioning for a week or more.	10%
3	Partially implemented	No policy exists. Some individual departments may have a "Shift without computer" practice where they turn off the monitor and try to do their job on paper. Some departments do not. The plant operators have a procedure to operate in manual. It has not been practiced or updated in years.	25%
4	Fully implemented and maintained	There is a policy written and reviewed annually that states the utility has a once every two years tabletop exercise. Where possible with minimal impact to real operations, power down computers, pull from inventory, and restore from offline backups. There are documented procedures on how to do things on paper or manually until the computer is restored.	60%

How to Manually Calculate Overall Vulnerability

The AWWA Tool evaluation should be about 100 questions/controls. This three-question analysis is just used to demonstrate how to calculate Detect, Delay, Respond, and overall vulnerability. See the example for a full evaluation.

Calculate the Sum of Scores and the Best Possible scores. In this demonstration, the numbers add up as shown in Table B.11.

Divide the Sum of Scores by the Best Possible to calculate the Percent Implemented for Detect, Delay, and Respond (Table B.12).

Ranges within Percent Implemented are used to select one of the four scores for Detect, Delay, and Respond.

Table B.10   Summary table of the demonstration example

Control Code	Description	Priority	Status	Explanation/ Example	Detect	Delay	Respond
IA-01	Access control policies and procedures established including unique user ID for every user, appropriate passwords, privilege accounts, authentication, and management oversight.	1 = 0.60	1 = 0.05 2 = 0.10 3 = 0.25 4 = 0.60	Based on their knowledge of access control policies, operators do not share passwords.	0	1 = 0.0300 2 = 0.0600 3 = 0.1500 4 = 0.3600	0
PE-02	Secure areas protected by entry controls and procedures to ensure that only authorized personnel have access.	1 = 0.60	1 = 0.05 2 = 0.10 3 = 0.25 4 = 0.60	Access to the server room is restricted to authorized staff only.	1 = 0.0300 2 = 0.0600 3 = 0.1500 4 = 0.3600	1 = 0.0300 2 = 0.0600 3 = 0.1500 4 = 0.3600	0
AU-05	Risk-based business continuity framework established under the auspices of the executive team to maintain continuity of operations and consistency of policies and plans throughout the organization. Another purpose of the framework is to ensure consistency across plans in terms of priorities, contact data, testing, and maintenance.	2 = 0.25	1 = 0.05 2 = 0.10 3 = 0.25 4 = 0.60	The facility has a documented and tested contingency plan to operate the facility without the use of SCADA software, in the case of attack by ransomware.	0	0	1 = 0.0125 2 = 0.0250 3 = 0.0625 4 = 0.1500

Table B.11   Method example—sum of scores and best possible scores

	Detect	Delay	Respond
Sum of Scores	0.3600	0.15 + 0.36 = 0.51	0.0625
Best Possible	0.3600	0.36 + 0.36 = 0.72	0.1500

Table B.12   Method example—percent implemented math

	Detect	Delay	Respond
Sum of Scores	0.3600	0.5100	0.0625
Best Possible	0.3600	0.7200	0.1500
Percent Implemented	0.36 / 0.36 = **1.00**	0.51 / 0.72 = **0.71**	0.625 / 0.15 = **0.42**

The three selected values for this demonstration are bold and italics in Table B.13.

Multiplying the three selected scores with each other will create the Combined Countermeasure Capability.

Combined Countermeasure Capability = (Certain – 4) × (Strong – 4) × (Slow – 1)

Combined Countermeasure Capability = (4 × 4 × 1)

Combined Countermeasure Capability = 16

Find the Combined Countermeasure Capability in Table B.14 to select the Overall Vulnerability.

The Combined Countermeasure of 16 converts to an Overall Vulnerability of 0.30 in this demonstration.

Table B.13 Method example—detect, delay, and respond weights and selection

Percent Implemented	Detect	Delay	Respond
0.7500–1.0000	*Certain – 4*	Very Strong – 6	Fast – 4
0.5000–0.7499	Probable – 2	*Strong – 4*	Variable – 2
0.2500–0.4999	Possible – 1	Limited – 2	*Slow – 1*
0.0000–0.2499	None – 0	None – 1	None – 0

Table B.14 Method combined countermeasure capability

Combined Countermeasure Capability	Capability Level	Overall Vulnerability
0	Low	1.00
1	Low	0.90
2	Low	0.80
4	Moderate	0.75
6	Moderate	0.60
8	Moderate	0.50
12	Moderate	0.40
16	*High*	*0.30*
24	High	0.20
32	High	0.12
48	Very high	0.09
64	Very high	0.06
96	Very high	0.03

Final Manual Instructions

It is envisioned that the evaluator would hand mark or circle the scores in the Answer Value (1–4) column and in the Detect/Delay/Respond columns. If a question/control does not apply, the evaluator would write N/A or cross though that control.

SECTION B.5: CYBERSECURITY THREAT LIKELIHOOD

Introduction

The likelihood time frame for cyber should be the same time frame used for natural and physical hazards. Typically, this is the likelihood of an attack in any one year. The maximum possible is one occurrence per year that becomes a 1.0, i.e., the attack will happen. The USEPA document EPA 817-F21-004 "Baseline Information on Malevolent Acts for Community Water Systems Version 2.0" released February 2021 has likelihood values for enterprise and SCADA systems. The USEPA estimate for a cyberattack on Business Enterprise Systems likelihood is 1.0 or once every 1 year or more frequently. The USEPA estimate for cyberattack on the PCS likelihood is 1.0 or once every 1 year or more frequently. Check the USEPA website for potentially updated numbers.

Based on the real-world experience of water and wastewater utilities, AWWA recommends a threat likelihood of 1.0 for all cyber threats (e.g., cyberattacks are attempted once per year or more often). Many utilities are under daily attack by cyber adversaries. Ransomware may not be released on the enterprise network every year. This does not mean that an attack was not attempted. It means that the organization protection systems are working against this threat. The credit for these protection systems is captured in the vulnerability score.

An example for comparison is a hurricane in Florida. The buildings may be elevated or hardened to withstand the hurricane, and there is minimal damage. The hurricanes happen once per year or more frequently. The defenses for hurricanes are stronger, vulnerability is lower, and risk is reduced. The likelihood of a hurricane in Florida is once per year or more frequently. The likelihood of a cyberattack is once per year or more frequently.

Cyberattack (Ransomware)—Business Enterprise Systems (IT)

Many organizations chose a likelihood of 1.0 occurrence per year for several reasons including but not limited to:

- Internet facing firewall is scanned multiple times daily.
- Malicious emails with damaging attachments or links to malicious websites make it through the filters. Employees click a link or open an infected file at least once a year. The local antimalware software catches it quickly, and it does not spread.

Cyber Theft or Diversion of the Financial System

Receive input from Finance or Human Resources for how often malevolent requests are received via email, text, or phone to perform any of the following actions:

- Change the Automated Clearing House or wire transfer instructions for a vendor payment.
- Change the direct deposit information of an employee.
- Receive a request to use money to purchase gift cards and send to the adversary.

If any combination of these occurs once per year or more frequently, a likelihood of 1.0 should be selected. If these are less-frequent occurrences, add all the occurrences of these over the time frame.

Cyberattack on the Process Control Systems

The likelihood of 1.0 for a cyberattack on the process control system is recommended.

Cyberattack on PCS—Sabotage Operator Changes

The likelihood of 1.0 for a cyberattack on the process control system is recommended.

Cyberattack on PCS—Sabotage Programming Changes

The likelihood of 1.0 for a cyberattack on the process control system is recommended.

Some may argue that this level of effort for sabotage is greater and happens less often than scanning of firewalls or release of malware to the environment. The USEPA recommendation of sabotage in 2019 is 0.05 or once every twenty years. However, there are incidents where remote attackers were able to gain remote access to the IT network. They were successful in pivoting to the PCS DMZ then to the PCS network. They were successful in elevating their privileges from operator to programmer. They were unable to change the programming because the PLC key was in run and not in program.

PCS attacks at the PLC level are attempted at least once per year. A value of 1.0 is recommended. A value of less than 0.05 is strongly discouraged.

SECTION B.6:   EXAMPLE CYBER ASSESSMENT

This example is provided as a detailed guideline of a fictitious utility. The vulnerability final values are provided during the discussion. The vulnerability interview answers are in Table B.16 followed by the tables of how the interview answers were converted to a final vulnerability score. Your utility should print out Table B.20 and circle your answers then do the math for your organization.

A water utility has a 15 MGD water treatment plant fed by a regional raw water provider. A distribution system of 31,500 water meters provides water to a population of 72,000 citizens. The terrain is somewhat hilly, and the distribution system has three pressure zones with a total of three elevated tanks and related pumping stations and three ground water wells. There is a single wastewater treatment plant with lift stations throughout the service area.

There is a new groundwater well being added to the distribution system. The construction is $3,000,000 and is two months into a 12-month construction plan. Raw water costs $150 per ac-ft that is paid quarterly.

The water utility charges an average of $4/1,000 gal. Assume each person averages 100 gal/day. That equates to 72,000 population × 100 gallons per day = 7.2 million gallons per day or 2.63 billion gallons per year—$10 million per year in water revenue that is controlled in an enterprise fund. New construction impact fees are not included in this example. The utility is part of a municipality that provides police, fire, street maintenance, trash pick-up, traffic control, city parks, bus service, and a library. The revenue for the city is $80 million per year from tax revenue and other fees. The combined income for the city is $90 million per year ($80M from tax and $10M from utility fees). There are 450 municipal employees, including 50 utility employees.

The water utility reads its own water meters and processes the meter reads and creates virtual invoices with a billing system. The invoices are processed and mailed by a third party. Utility customer service representatives accept cash and check payments at city hall. The city customer service supervisor makes daily deposits.

A third-party processes online payment, provides a "payment received" file to be imported to the billing system, and transfers funds to the bank. The third party processes all credit card transactions even if presented in person.

The IT department supports the on-premises billing and financial servers along with other typical utility services including GIS and work order systems. There are 75 servers and 450 computers on the IT network.

The IT department provides networking services for IT and PCS. The water utility PCS system has a dedicated VLAN of the IT network at the treatment facilities. There is a dedicated PCS support team that supports the HMI software, PLCs, sensor calibrations, and unlicensed radios to the field assets. There is a pair of redundant PCS servers at each treatment plant. There are eight PCS clients at each facility. There are multiple PLC cabinets with local indications and control push buttons at each location.

The assets selected are:
- Finance system
- Enterprise IT system
- PCS system

The selected threat-asset pairs selected are:

- Cyber theft or diversion of the financial system
- Cyberattack on the business enterprise systems (IT)—ransomware
- Cyberattack on the process control systems—ransomware
- Cyberattack on the process control systems—sabotage operator changes
- Cyberattack on the process control systems—sabotage programmer changes

The AWWA tool is selected to evaluate vulnerability. The tool is run twice—once for the IT network and once for the PCS because of the network separation and the different support teams. The IT network team attends the PCS session.

Vulnerability Calculations

See Table B.16 for the responses to the IT interview, the PCS interview, and vulnerability calculations.

The IT vulnerability is calculated at 0.75 with the AWWA Manual Tool.

The PCS vulnerability is calculated at 0.80 with the AWWA Manual Tool.

The assessment team agrees with these numbers.

Cyber Theft or Diversion of the Financial System

Consequence: Cyber Theft or Diversion of the Financial System.

There are a couple of options to consider between the well site construction progress payment and the raw water supply payment. Showing the math for both considerations to select the highest consequence option.

Consequence Option 1. The well site being built has an average monthly progress payment of the total purchase/time period.

$3,000,000/12 months = $250,000 per monthly payment.

Consequence Option 2. The raw water bill is 2.63 billion gallons per year/(325,851 gallons per ac-ft) × ($150 per ac-ft)/4 quarters per year = $302,669 per quarterly payment.

Consequence option 2 is the larger value.

The consequence is $302,669.

Vulnerability—Cyber Theft or Diversion of the Financial System

The Chief Financial Officer attends the IT interview and indicates that there is dedicated mandatory documented training for the customer service and finance staff discussing business email compromise. There are documented processes mandating that changing the banking information for individuals or companies shall be verified with a different communication method.

The AWWA tool could be run again for this threat–asset pair. The IT vulnerability assessment was 0.75 with the manual AWWA tool. The financial system is given 0.15 for additional training by the assessment team. The team agrees to the vulnerability for the "Finance System—Cyber Theft" is 0.60.

Likelihood—Cyber Theft or Diversion of the Financial System

During the IT interview, the team shared that there have been approximately 4 times in the past year where an adversary has texted impersonating the CEO asking for gift cards to be sent or an email asking for an immediate wire transfer to a bank out of the country.

The likelihood selected is 1.0.

Risk—Cyber Theft or Diversion of the Financial System

Risk is Consequence × Vulnerability × Likelihood.

Risk = $302,669 × 0.60 × 1.0

Risk = $181,601

Annual risk for cyber theft or diversion of the financial system = **$181,601**

Cyberattack on the Business Enterprise Systems (IT)—Ransomware

Consequence: Cyberattack on the water business enterprise systems (IT)—ransomware. Enterprise IT indicates that they use online nightly incremental backups and weekly full backups that they automatically move to cloud storage. They have restored accidentally deleted files from the online backups. They have never tried to restore a server from a backup in the cloud. They do not have any standing contracts for resources to help rebuild after a ransomware attack.

The water utility PCS and power generation PCS are assumed to be unaffected.

Method 1 was not selected because the city has not contemplated the cost of a ransomware attack. The city does not have cyber insurance and would be responding to the attack ad hoc. The evaluation team selected Method 2.

Using Method 2, the CISA estimates that the ransomware attack will cost 0.80% of annual revenue.

Annual revenue is $90 million. The estimated ransomware attack will cost the entire city:

$90,000,000 × 0.008 = $720,000

Water has about 11% of the staff of the city (50 of 450 employees). The water portion of the consequence is 50/450 of the entire city consequence:

Consequence = $720,000 × 50/450

Consequence = $80,000

Vulnerability—Cyberattack on the Business Enterprise Systems (IT)—Ransomware

The IT vulnerability is calculated by the AWWA Tool; documented is 0.75 (See Table B.19).

Likelihood—Cyberattack on the Business Enterprise Systems (IT)—Ransomware

The enterprise IT team indicates that the antimalware software indicates that malware is found once per year or more often. A value of 1 occurrence per year is agreed to. This correlates to a likelihood of 1.00.

Risk–Cyberattack on the Business Enterprise Systems (IT)—Ransomware

Risk = Consequence × Vulnerability × Likelihood

Risk = $80,000 × 0.75 × 1.0

Risk = $60,000

Annual risk for a cyberattack on the water business enterprise systems (IT) – ransomware is **$60,000.**

Cyberattack on the Process Control Systems–Ransomware

Consequence: Cyberattack on the process control systems— ransomware. This exercise is to calculate the cost consequence on the PCS side of a ransomware attack.

Assume:

- There is a PCS support team of three individuals including the PCS manager.

- Both the water treatment plant and wastewater treatment plant are impacted.

- There is a good relationship between the municipality and the engineering company that designed and programmed the wastewater treatment plant. They provide three consultants to help recover the plant they are familiar with. It takes 18 hours to get them on site.

- The PCS support team has backups of the HMI servers made a month ago on a portable drive kept in the PCS support team manager's desk.

- There is a PCS support laptop for the water side that has the PLC programs. The laptop was powered down during the attack.

- There is a different PCS support laptop for the wastewater side that has the PLC programs. The laptop was powered down during the attack.

- Each plant typically has 6 individuals assigned to cover 24/365 operation.

- There is an additional person that visits each remote facility daily and grabs water quality samples to deliver to the lab. This is a total complement of 13 operators.

- The average loaded salary (with benefits) is $45 per hour for municipal employees.

- The average billing rate for engineering consultants with PCS capabilities is $200 per hour.

Use the assumption that it will take the SCADA support team 75 hours (3 calendar days) to restore the PCS servers and the clients at each facility. A fourth day at these staffing levels to watch and verify operations.

- SCADA Consequence = 3 SCADA Support Team for 4 days and 8 hours per day at $45/hour x 1.5 for overtime or weekends
- SCADA Consequence = $3 \times 4 \times 8 \times 45 \times 1.5$
- SCADA Consequence = $6,480

Water treatment plant: 21 plant operators and 6 rovers working 8-hour shifts around the clock to restore the system and do point-to-point checks for 4 days.

- Water Consequence = 27 employees $\times$ 8 h per day $\times$ 4 days $\times$ $45/ h $\times$ 1.5 for overtime or weekends
- Water Consequence = $27 \times 8 \times 4 \times 45 \times 1.5$
- Water Consequence = $58,320

The wastewater plant will be waiting the first day for the consultants to arrive. Manual operation is needed for 5 days. The support team is:

- WW Consultants Consequence = 3 consultants for 5 days and 8 h per day at $200/h
- WW Consultants Consequence = $3 \times 5 \times 8 \times 200$
- WW Consultants Consequence = $24,000
- WW Consequence = 24 employees $\times$ 8 h per day $\times$ 5 days $\times$ $45/h $\times$ 1.5 for overtime or weekends
- WW Consequence = $24 \times 8 \times 5 \times 45 \times 1.5$
- WW Consequence = $64,800

The Emergency Operations Center and coordination with upper management and outside agencies will be needed. Assumed these costs to be 100% to these estimates and multiplied the costs by 2.

- EOC Consequence = Recovery Costs $\times$ 2

All Consequences

- Consequence = SCADA Consequence + Water Consequence + WW Consultants Consequence + WW Consequence + EOC Consequence
- Consequence = ($6,480 + $58,320 + $ 24,000 + $ 64,800) $\times$ 2
- Consequence = **$307,200**

106

Vulnerability—Cyberattack on the Process Control Systems—Ransomware

The PCS vulnerability calculated by the AWWA Tool is 0.80 (See Table B.19).

Likelihood—Cyberattack on the Process Control Systems—Ransomware

The PCS support team indicates that the antimalware software has never found any malicious software on the PCS. Remote access to the SCADA system is allowed by VPN to the enterprise network. If malicious software was found on the IT network, it would likely spread to the PCS. A value of 1 occurrence per year is agreed to. This correlates to a likelihood of 1.0.

Risk—Cyberattack on the Process Control Systems—Ransomware

Risk is Consequence × Vulnerability × Threat Likelihood

Risk = $307,200 × 0.80 × 1.0

Risk = $245,760

Annual risk for a cyberattack on the process control systems – ransomware = **$245,760**

Cyberattack on the Process Control Systems—Sabotage Operator Changes

Assume that an adversary was able to remote into the PCS and turned the caustic (sodium hydroxide) chemical feed to maximum. The plant operator tours the facility every 4 h making rounds including the chemical feed pumps. There is a 250,000-gal clear well at the discharge of the plant that has a pH sensor. There is no automatic isolation of the discharge valve from the clear well based on pH. Plant transit time is approximately 11 h at rated capacity.

The operator notices the chemical pump is running significantly faster than it was 4 h ago during the last set of rounds. The operator used the PCS trending to verify that there was a step change in chemical addition approximately 3 h ago. The operator puts the chemical addition rate back to the previous setting and contacts the supervisor. They confirm that none of the qualified operators made the change. They inform the Utility Directory. The Utility Director contacts the Chief of Police and City Manager.

The water superintendent instructs the operator to start up the distribution wells and place the plant offline. Water quality samples at the discharge of the plant and the distribution system confirm good water quality. The well sites are not able to keep up with demand, and the elevated tank levels are falling. There are approximately eight hours to get the plant back online.

The plant operators call in help to drain the filters and clear well. In an overabundance of caution, the city manager decides to notify the public that the treatment plant experienced a cyberattack, and chemical feed pumps were set at maximum. Only safe drinking water left the treatment plant, but the treatment plant is offline until the issues are resolved. Please conserve water by not watering lawns and washing cars for the next 24 hours. There is a media frenzy, and it makes the national news.

Consequence—Cyberattack on the Process Control Systems—Sabotage Operator Changes

The estimated time and effort to drain the plant is $100,000.

The estimated effort from upper management and media support is $100,000.

Total consequence is $100,000 + $100,000

Consequence = $200,000

Likelihood—Cyberattack on the Process Control Systems—Sabotage Operator Changes

There are a few recorded attempts of this occurring at water agencies in the United States. Attempted remote access attacks on PCS are getting more frequent. The assessment team agrees to a recommended likelihood of 1.0 or one every 1 year.

Risk—Cyberattack on the Process Control Systems—Sabotage Operator Changes

Risk is Consequence × Vulnerability × Threat Likelihood

Risk = $200,000 × 0.8 × 1.0

Risk = $160,000

Annual risk for a cyberattack on the process control systems – sabotage operator changes = **$160,000.**

Cyberattack on the Process Control Systems—Sabotage Programmer Changes

An advanced adversary remotely gained access to the PCS network undetected for a significant amount of time. The PLCs do not have passwords. The adversary can download the programs and work on them offline. The adversary has access to all the PLCs and configures all the distribution system pumps to start at the same time and ramp up to maximum speed including the well sites that close the waste valves to push the water to the distribution system.

The distribution system has three pressure zones. Two pressure zones have reservoirs that fill and drain as demand changes. The pumping will cause these reservoirs to go up and overflow. There will be a raise in pressure, but main breaks are not expected. These two pressure zones provide water to approximately 75% of the population.

The third pressure zone provides water to approximately 25% of the population. It is a closed system, and pressure in the system is maintained by the speed and quantity of the pumps. When these pumps go to full speed, there is significant damage expected to this system.

Consequence—Cyberattack on the Process Control Systems—Sabotage Programming Changes

It is decided to use the WHEAT calculator for this consequence.

Utility Information:

- Utility name – Testing
- State – Nevada
- Zip code – 89044 [Area 51]
- Population served – 75,000
- Ownership – Public
- Average daily water service (MGD) – 8
- Avg. rate ($/1,000 gal) – $4.00

Utility Resilience Index – Skipped for this exercise.

Qualitative Risk Assessment

Storage and Distribution Facilities – Malevolent Act

- Pump Discharge Main
- Cyberattack on Process Control Systems

Click WHEAT Calculator

- Loss of Operating Asset
- Finished Water Distribution Main
- Enter length of pipe affected in feet: 1,000 ft.

Click Save and Continue

- Cost – $1,000,000
- Duration of Service Outage (Days) – 14
- Customers without service (%) – 25%

The calculated impacts are:

- Fatalities – 0
- Injuries – 0
- Utility Financial Impact – $133,000
- Regional Economic Consequences – $78,339,000

The Regional Economic Consequences were chosen to be ignored. There could be some justification for including the Regional Economic Consequences.

Total consequence = **$133,000**

Likelihood—Cyberattack on the PCS Equipment—Sabotage Programming Changes

Since the Regional Economic Consequences were ignored, the assessment group agrees to likelihood of 1.0 or 1 occurrence every 1 year. If the Regional Economic Consequences were included, a likelihood of 0.001 may be selected.

Vulnerability—Cyberattack on the PCS Equipment— Sabotage Programming Changes

Vulnerability from the PCS assessment is 0.80 (See Table B.19).

Risk–Cyberattack on the PCS Equipment–Sabotage Programming Changes

Risk is Consequence × Vulnerability × Threat Likelihood

Risk = $133,000 × 0.80 × 1.0

Risk = **$106,400**

Annual risk for cyberattack on the PCS equipment – sabotage programming changes = **$106,400**

Table B.15 Risk assessment example summary

Asset	Threat	Risk	Consequence	Vulnerability	Likelihood
Finance System	Cyber theft	$181,601	$302,669	0.60	1.0
Enterprise IT	Cyberattack on business enterprise/ IT network— ransomware	$60,000	$80,000	0.75	1.0
PCS	Cyberattack on the PCS—ransomware	$245,760	$307,200	0.80	1.0
PCS	PCS equipment— sabotage operator access	$160,000	$200,000	0.80	1.0
PCS	PCS equipment— sabotage programmer access	$106,400	$133,000	0.80	1.0

Vulnerability Calculations for the Example

Table B.16 is used for the example. It is a copy of the selections from Table B.20 with answers to the list questions/control measures in the AWWA CRAT version 3.0 as of September 2021. In the interest of saving paper, the IT and PCS evaluations are presented in Table B.16. Questions/Controls that are not applicable are shown as a dash (-).

Column Explanations

Control: The number of the control. Leading zeros were added to single digit control.

Priority: The priority of the control from the AWWA CRAT

Status IT: The assessor's determination of the status of that control during the IT assessment

Status PCS: The assessor's determination of the status of that control during the PCS assessment

DDR IT – Detect/Delay/Respond score for the IT assessment. Not all controls have D, D, and R.

DDR PCS – Detect/Delay/Respond score for the PCS assessment. Not all controls have D, D, and R.

This ends the interview portion of the example assessment.

Table B.16 Cyber example vulnerability answers

Control	Priority	Status IT	DDR IT	Status PCS	DDR PCS
AT-03	1	0.05	0.0300	0.05	0.0300
AU-01	1	0.10	0.0600	0.10	0.0600
AU-02	1	0.25	0.1500	0.10	0.0600
AU-03	1	0.25	0.1500	0.10	0.0600
AU-04	1	0.60	0.3600	0.10	0.0600
DS-01	1	–	–	–	–
DS-02	1	0.10	0.0600	–	–
DS-03	1	–	–	–	–
IA-01	1	0.60	0.3600	0.10	0.0600
IA-10	1	0.60	0.3600	0.10	0.0600
IA-11	1	0.60	0.3600	0.60	0.3600
IA-12	1	0.60	0.3600	0.60	0.3600
IA-03	1	0.25	0.1500	0.10	0.0600
IA-04	1	0.10	0.0600	0.10	0.0600
IA-05	1	0.60	0.3600	0.60	0.3600
IA-06	1	0.10	0.0600	0.10	0.0600
IA-07	1	0.25	0.1500	–	–
IA-09	1	0.60	0.3600	0.60	0.3600
PE-01	1	0.25	0.1500	0.25	0.1500
PE-02	1	0.60	0.3600	0.05	0.0300
PE-03	1	0.25	0.1500	0.25	0.1500
PE-04	1	0.60	0.3600	0.60	0.3600
PE-05	1	0.10	0.0600	0.10	0.0600
PE-06	1	0.25	0.1500	0.25	0.1500
PE-07	1	0.25	0.1500	0.25	0.1500
PE-08	1	0.60	0.3600	0.25	0.1500
PE-09	1	0.60	0.3600	0.60	0.3600
SC-01	1	0.25	0.1500	0.10	0.0600
SC-11	1	–	–	–	–
SC-12	1	0.25	0.1500	0.25	0.1500
SC-14	1	0.25	0.1500	0.25	0.1500
SC-15	1	0.25	0.1500	0.25	0.1500
SC-16	1	0.25	0.1500	0.25	0.1500
SC-17	1	0.60	0.3600	0.60	0.3600
SC-18	1	0.25	0.1500	–	–
SC-19	1	0.60	0.3600	–	–
SC-02	1	0.60	0.3600	0.05	0.0300
SC-20	1	–	–	0.05	0.0300

(continued)

Table B.16 Cyber example vulnerability answers (*Continued*)

Control	Priority	Status IT	DDR IT	Status PCS	DDR PCS
SC-21	1	0.60	0.3600	–	–
SC-22	1	0.05	0.0300	–	–
SC-23	1	0.60	0.3600	0.60	0.3600
SC-24	1	0.05	0.0300	0.05	0.0300
SC-25	1	0.60	0.3600	0.60	0.3600
SC-03	1	0.10	0.0600	0.10	0.0600
SI-01	1	0.25	0.1500	–	–
SI-03	1	0.60	0.3600	0.05	0.0300
AT-01	2	0.10	0.0250	0.10	0.0250
AT-02	2	0.10	0.0250	0.10	0.0250
AU-05	2	0.25	0.0625	0.25	0.0625
AU-06	2	0.10	0.0250	0.10	0.0250
AU-07	2	–	–	–	–
CM-03	2	0.25	0.0625	0.10	0.0250
CM-04	2	0.10	0.0250	0.05	0.0125
CM-05	2	0.10	0.0250	0.10	0.0250
CM-07	2	0.25	0.0625	0.10	0.0250
IR-01	2	0.10	0.0250	0.10	0.0250
MA-03	2	0.10	0.0250	0.10	0.0250
PM-03	2	0.25	0.0625	0.10	0.0250
PM-04	2	0.10	0.0250	0.10	0.0250
RA-01	2	0.25	0.0625	0.25	0.0625
RA-02	2	0.25	0.0625	0.25	0.0625
SC-10	2	0.25	0.0625	0.10	0.0250
SC-13	2	0.25	0.0625	0.25	0.0625
SC-04	2	0.10	0.0250	0.10	0.0250
SC-05	2	0.10	0.0250	0.10	0.0250
SC-06	2	0.25	0.0625	0.05	0.0125
SC-07	2	0.10	0.0250	–	–
SC-08	2	0.10	0.0250	0.10	0.0250
SC-09	2	0.10	0.0250	0.10	0.0250
SI-02	2	0.25	0.0625	0.05	0.0125
SI-05	2	0.25	0.0625	0.25	0.0625
AU-08	3	0.10	0.0100	0.10	0.0100
CIE-01	3	–	–	0.25	0.0250
CM-01	3	0.25	0.0250	0.25	0.0250
CM-02	3	0.25	0.0250	0.25	0.0250
IA-08	3	0.25	0.0250	0.25	0.0250

(*continued*)

Table B.16 Cyber example vulnerability answers (*Continued*)

Control	Priority	Status IT	DDR IT	Status PCS	DDR PCS
IR-03	3	0.10	0.0100	0.10	0.0100
MP-02	3	0.10	0.0100	0.10	0.0100
PM-01	3	0.25	0.0250	0.25	0.0250
PM-02	3	0.25	0.0250	0.25	0.0250
PS-01	3	0.60	0.0600	0.60	0.0600
PS-02	3	0.10	0.0100	0.10	0.0100
PS-03	3	0.10	0.0100	0.10	0.0100
PS-04	3	0.25	0.0250	0.25	0.0250
SA-01	3	0.25	0.0250	0.05	0.0050
SA-02	3	0.25	0.0250	0.05	0.0050
SA-03	3	0.25	0.0250	0.05	0.0050
SA-04	3	0.25	0.0250	0.25	0.0250
SA-05	3	0.10	0.0100	0.10	0.0100
SI-04	3	0.60	0.0600	0.25	0.0250
CM-06	4	0.10	0.0050	0.10	0.0050
IA-02	4	0.60	0.0300	0.60	0.0300
IR-02	4	0.10	0.0050	0.10	0.0050
MA-01	4	0.10	0.0050	0.10	0.0050
MA-02	4	0.10	0.0050	0.10	0.0050
MP-01	4	0.10	0.0050	0.10	0.0050
MP-03	4	0.60	0.0300	0.60	0.0300
PM-05	4	0.05	0.0025	0.05	0.0025
SU-01	3	0.05	0.0050	0.05	0.0050
SU-02	4	0.05	0.0025	0.05	0.0025

Table B.17 Sum of vulnerability answers and sum of max possible

	IT Assessment			PCS Assessment		
	Detect	Delay	Respond	Detect	Delay	Respond
Sum of Selected	3.2375	9.4225	2.2525	2.19	5.8125	1.5025
Sum of Max Possible	6.8400	16.2300	4.8600	6.9000	13.9800	4.5000
Sum of Selected Divided by Sum of Max Possible	0.4733	0.5806	0.4635	0.3174	0.4158	0.3339

Table B.18 Convert percentages to score

Score Range	IT Assessment Scores			PCS Assessment Scores		
	Detect	Delay	Respond	Detect	Delay	Respond
0.7500– 1.000	Certain - 4	Very Strong - 6	Fast - 4	Certain - 4	Very Strong - 6	Fast - 4
0.5000– 0.7499	Probable - 2	Strong - 4	Variable - 2	Probable - 2	Strong - 4	Variable - 2
0.2500– 0.4999	Possible - 1	Limited - 2	Slow - 1	Possible - 1	Limited - 2	Slow - 1
0.0000– 0.24999	None - 0	None - 1	None - 0	None - 0	None - 1	None - 0

Table B.19 Combined reference table

Combined Reference	Capability Level	Vulnerability	
0	Low	1.00	
1	Low	0.90	
2	Low	0.80	<PCS
4	Moderate	0.75	<IT
6	Moderate	0.60	
8	Moderate	0.50	
12	Moderate	0.40	
16	High	0.30	
24	High	0.20	
32	High	0.12	
48	Very High	0.09	
64	Very High	0.06	
96	Very High	0.03	

To determine vulnerability from the answers provided, follow this five-step process:

(1) Add up all the selected scores for Detect, then Delay, then Respond.

(2) Add up the max possible for Detect, then Delay, then Respond accounting for the not applicable controls.

(3) Take the sums and divide them by the sum of the maximum possible scores (Table B.17).

(4) Select Detect, Delay, and Respond based on the Score Range from Table B.18.

(5) Multiply the scores together to get the combined score.

IT Vulnerability Assessment = Possible – 1 × Strong – 4 × Slow – 1 = **4**

PCS Vulnerability Assessment = Possible – 1 × Limited – 2 × Slow – 1 = **2**

(6) Based on these answers, select the vulnerability from Table B.19.

The vulnerability for the IT Assessment and the related threat–asset pairs is 0.75.

The vulnerability for the PCS Assessment and the related threat–asset pairs is 0.80.

SECTION B.7: MANUAL VULNERABILITY ASSESSMENT WORKSHEET

This table is a list of the questions/control measures in the AWWA Cybersecurity Risk Assessment Tool (CRAT) version 3.0 as of September 2021 (see Sec. B.8 References). The Detect, Delay, or Respond columns indicate if the control measure has an impact on any combination of Detect, Delay, or Respond. For example, the first entry on Table B.15 is AT-03 and describes a forensics program. A forensics program will not help with detection or with delay. It only helps with Respond and therefore is only in the Respond column.

The Priority value comes from the AWWA CRAT. Multiplying Priority * Status will generate the values in detect, delay and/or respond. The math has been done, and the assessor circles the single status or the detect, delay and/or respond line of values.

To manually perform the assessment, print out this appendix. Print out this section. Circle the appropriate single status of 1, 2, 3, or 4. Circle the matching value in Detect, Delay, and/or Respond. If this question or control does not apply because that technology is not used, then cross out that row. The Control Codes presented have a leading 0 added to the number. The control codes in AWWA CRAT Online do not have the leading zeros. Separate assessments can be done for IT and SCADA. The intent is the auditor will print out the table and circle the selection in the Status column during the interview process.

See the instructions at the end of the questions/controls to manually calculate vulnerability.

Table B.20 Manual vulnerability assessment table

Control Code	Description	Priority	Status	Explanation/Example	Detect	Delay	Respond
AT-03	A forensic program established to ensure that evidence is collected/handled in accordance with pertinent laws in case of an incident requiring civil or criminal action.	1 = 0.60	1 = 0.05 2 = 0.10 3 = 0.25 4 = 0.60	A SCADA tech believes a machine is infected. Based on their training, they remove the machine from the network and report it to IT without powering it off to avoid deleting evidence.	0	0	1 = 0.0300 2 = 0.0600 3 = 0.1500 4 = 0.3600
AU-01	Audit program established to ensure information systems are compliant with policies and standards and to minimize disruption of operations.	1 = 0.60	1 = 0.05 2 = 0.10 3 = 0.25 4 = 0.60	IT schedules an independent review and examination of records and activities to assess the adequacy of system controls and to ensure compliance with established policies.	1 = 0.0300 2 = 0.0600 3 = 0.1500 4 = 0.3600	1 = 0.0300 2 = 0.0600 3 = 0.1500 4 = 0.3600	1 = 0.0300 2 = 0.0600 3 = 0.1500 4 = 0.3600
AU-02	Framework of information security policies, procedures, and controls including management's initial and periodic approval established to provide governance, and exercise periodic review, dissemination, and coordination of information security activities.	1 = 0.60	1 = 0.05 2 = 0.10 3 = 0.25 4 = 0.60	The process of implementing policies and procedures is clearly defined and reviewed. Updates to this process are made by a responsible party.	1 = 0.0300 2 = 0.0600 3 = 0.1500 4 = 0.3600	1 = 0.0300 2 = 0.0600 3 = 0.1500 4 = 0.3600	1 = 0.0300 2 = 0.0600 3 = 0.1500 4 = 0.3600
AU-03	Governance framework to disseminate/decentralize decision-making while maintaining executive authority and strategic control and ensure that managers follow the security policies and enforce the execution of security procedures within their area of responsibility.	1 = 0.60	1 = 0.05 2 = 0.10 3 = 0.25 4 = 0.60	A utility has a required number of accountable staff that must review or provide input before security policies are put in place. Periodic review that approved security policies are being followed.	1 = 0.0300 2 = 0.0600 3 = 0.1500 4 = 0.3600	1 = 0.0300 2 = 0.0600 3 = 0.1500 4 = 0.3600	1 = 0.0300 2 = 0.0600 3 = 0.1500 4 = 0.3600
AU-04	Information security responsibilities defined and assigned.	1 = 0.60	1 = 0.05 2 = 0.10 3 = 0.25 4 = 0.60	All staff are aware of who they would report to if they notice suspicious behavior in the system.	0	0	1 = 0.0300 2 = 0.0600 3 = 0.1500 4 = 0.3600
DS-01	A program established to ensure compliance with the minimum PCI requirements for your associated level.	1 = 0.60	1 = 0.05 2 = 0.10 3 = 0.25 4 = 0.60	The company selected to perform billing is compliant with the minimum PCI requirements for the utility's associated level.	0	1 = 0.0300 2 = 0.0600 3 = 0.1500 4 = 0.3600	0
DS-02	A Privacy Policy as well as a Cyber Security Breach Policy are implemented.	1 = 0.60	1 = 0.05 2 = 0.10 3 = 0.25 4 = 0.60	An operator knows how to identify and respond to a suspected cyber breach based on their cybersecurity training.	0	1 = 0.0300 2 = 0.0600 3 = 0.1500 4 = 0.3600	1 = 0.0300 2 = 0.0600 3 = 0.1500 4 = 0.3600

(continued)

117

Table B.20 Manual vulnerability assessment table *(Continued)*

Control Code	Description	Priority	Status	Explanation/Example	Detect	Delay	Respond
DS-03	A program is established to ensure compliance with the minimum HIPAA requirements. Develop a Privacy Policy as well as a Cyber Security Breach Policy.	1 = 0.60	1 = 0.05 2 = 0.10 3 = 0.25 4 = 0.60	Current practices are reviewed by legal counsel for legal compliance with HIPAA.	0	0	1 = 0.0300 2 = 0.0600 3 = 0.1500 4 = 0.3600
IA-01	Access control policies and procedures established including unique user ID for every user, appropriate passwords, privilege accounts, authentication, and management oversight.	1 = 0.60	1 = 0.05 2 = 0.10 3 = 0.25 4 = 0.60	Based on their knowledge of access control policies, operators do not share passwords.	0	1 = 0.0300 2 = 0.0600 3 = 0.1500 4 = 0.3600	0
IA-10	Policies and procedures for least privilege established to ensure that users only gain access to the authorized services.	1 = 0.60	1 = 0.05 2 = 0.10 3 = 0.25 4 = 0.60	If no user is logged in at a SCADA screen, a read-only view is presented. Individual roles created and assigned to users depending on their responsibilities.	1 = 0.0300 2 = 0.0600 3 = 0.1500 4 = 0.3600	1 = 0.0300 2 = 0.0600 3 = 0.1500 4 = 0.3600	0
IA-11	Workstation and other equipment authentication framework established to secure sensitive access from certain high-risk locations.	1 = 0.60	1 = 0.05 2 = 0.10 3 = 0.25 4 = 0.60	Access to control of critical equipment is only available at a secured terminal.	0	1 = 0.0300 2 = 0.0600 3 = 0.1500 4 = 0.3600	0
IA-12	Session controls established to inactivate idle sessions, provide web content filtering, prevent access to malware sites, etc.	1 = 0.60	1 = 0.05 2 = 0.10 3 = 0.25 4 = 0.60	An operator attempts to connect to a known hacking website. The connection is blocked. The operator and IT are notified of the attempt.	1 = 0.0300 2 = 0.0600 3 = 0.1500 4 = 0.3600	1 = 0.0300 2 = 0.0600 3 = 0.1500 4 = 0.3600	0
IA-03	Role-based access control system established, including policies and procedures.	1 = 0.60	1 = 0.05 2 = 0.10 3 = 0.25 4 = 0.60	SCADA software implements unique usernames and passwords with different levels of control based on roles.	0	1 = 0.0300 2 = 0.0600 3 = 0.1500 4 = 0.3600	0
IA-04	Access control for confidential system documentation established to prevent unauthorized access of trade secrets, program source code, documentation, and passwords (including approved policies and procedures).	1 = 0.60	1 = 0.05 2 = 0.10 3 = 0.25 4 = 0.60	Defined clearance requirements for individuals to access confidential information.	0	1 = 0.0300 2 = 0.0600 3 = 0.1500 4 = 0.3600	0

(continued)

118

Table B.20 Manual vulnerability assessment table *(Continued)*

Control Code	Description	Priority	Status	Explanation/Example	Detect	Delay	Respond
IA-05	Access control for diagnostic tools and resources and configuration ports.	1 = 0.60	1 = 0.05 2 = 0.10 3 = 0.25 4 = 0.60	PLC programming software is only available at select workstations and only accessible to SCADA technicians.	0	1 = 0.0300 2 = 0.0600 3 = 0.1500 4 = 0.3600	0
IA-06	Access control for networks shared with other parties in accordance with contracts, SLAs, and internal policies.	1 = 0.60	1 = 0.05 2 = 0.10 3 = 0.25 4 = 0.60	Contracts with third-party equipment vendors establish security requirements for remote access to equipment.	0	1 = 0.0300 2 = 0.0600 3 = 0.1500 4 = 0.3600	0
IA-07	Wireless and guest-access framework established for the management, monitoring, review, and audit of wireless and guest access in place.	1 = 0.60	1 = 0.05 2 = 0.10 3 = 0.25 4 = 0.60	To use the plant guest network, users are required to accept a user agreement.	0	1 = 0.0300 2 = 0.0600 3 = 0.1500 4 = 0.3600	0
IA-09	Multifactor authentication system established for critical areas.	1 = 0.60	1 = 0.05 2 = 0.10 3 = 0.25 4 = 0.60	Remote access to the SCADA system requires two-factor authentication.	0	1 = 0.0300 2 = 0.0600 3 = 0.1500 4 = 0.3600	0
PE-01	Security perimeters, card-controlled gates, manned booths, and procedures for entry control.	1 = 0.60	1 = 0.05 2 = 0.10 3 = 0.25 4 = 0.60	Personnel are required to present a badge to access the PCS.	0	1 = 0.0300 2 = 0.0600 3 = 0.1500 4 = 0.3600	0
PE-02	Secure areas protected by entry controls and procedures to ensure that only authorized personnel have access.	1 = 0.60	1 = 0.05 2 = 0.10 3 = 0.25 4 = 0.60	Access to the server room is restricted to authorized staff only.	1 = 0.0300 2 = 0.0600 3 = 0.1500 4 = 0.3600	1 = 0.0300 2 = 0.0600 3 = 0.1500 4 = 0.3600	0
PE-03	Physical security and procedures for offices, rooms, and facilities.	1 = 0.60	1 = 0.05 2 = 0.10 3 = 0.25 4 = 0.60	Staff lock doors that allow access to PCS assets. Security guards inspect doors to make sure they are locked properly.	0	1 = 0.0300 2 = 0.0600 3 = 0.1500 4 = 0.3600	0
PE-04	Physical protection against fire, flood, earthquake, explosion, civil unrest, etc.	1 = 0.60	1 = 0.05 2 = 0.10 3 = 0.25 4 = 0.60	Fire suppression unit installed around critical equipment.	0	1 = 0.0300 2 = 0.0600 3 = 0.1500 4 = 0.3600	1 = 0.0300 2 = 0.0600 3 = 0.1500 4 = 0.3600
PE-05	Physical security and procedures for working in secure areas.	1 = 0.60	1 = 0.05 2 = 0.10 3 = 0.25 4 = 0.60	Documentation for physical security procedures is included with new employee training and reviewed at regular training events.	0	1 = 0.0300 2 = 0.0600 3 = 0.1500 4 = 0.3600	0
PE-06	Physical security and procedures for mail rooms, loading areas, etc., established. These areas must be isolated from IT/ PCS areas.	1 = 0.60	1 = 0.05 2 = 0.10 3 = 0.25 4 = 0.60	Server room and PLC cabinets are isolated from areas that delivery personnel and customers may visit.	0	1 = 0.0300 2 = 0.0600 3 = 0.1500 4 = 0.3600	0

(continued)

Table B.20   Manual vulnerability assessment table *(Continued)*

Control Code	Description	Priority	Status	Explanation/Example	Detect	Delay	Respond
PE-07	Physical security and procedures against equipment environmental threats and hazards or unauthorized access.	1 = 0.60	1 = 0.05 2 = 0.10 3 = 0.25 4 = 0.60	The utility monitors facilities using security cameras.	1 = 0.0300 2 = 0.0600 3 = 0.1500 4 = 0.3600	0	0
PE-08	Physical/Logical protection against power failure of equipment (uninterruptible power supplies [UPS]).	1 = 0.60	1 = 0.05 2 = 0.10 3 = 0.25 4 = 0.60	UPS are available as power backup for critical components.	0	1 = 0.0300 2 = 0.0600 3 = 0.1500 4 = 0.3600	1 = 0.0300 2 = 0.0600 3 = 0.1500 4 = 0.3600
PE-09	Physical/Logical protection against access to power and telecommunications cabling established.	1 = 0.60	1 = 0.05 2 = 0.10 3 = 0.25 4 = 0.60	A utility has a standby power source with separated power cabling for critical sites.	0	1 = 0.0300 2 = 0.0600 3 = 0.1500 4 = 0.3600	1 = 0.0300 2 = 0.0600 3 = 0.1500 4 = 0.3600
SC-01	Policies and procedures governing cryptography and cryptographic protocols including key/certificate-management established to maximize protection of systems and information.	1 = 0.60	1 = 0.05 2 = 0.10 3 = 0.25 4 = 0.60	When selecting new PLCs for a system upgrade, SCADA techs evaluate the option of using newer PLCs that offer encryption for communication.	0	1 = 0.0300 2 = 0.0600 3 = 0.1500 4 = 0.3600	0
SC-11	Framework for hardening of mobile code and devices established (including acceptance criteria and approved policies and procedures).	1 = 0.60	1 = 0.05 2 = 0.10 3 = 0.25 4 = 0.60	A water utility chooses to not allow personal mobile devices to connect to the control network. The utility does provide mobile devices managed by IT that can connect to the network.	1 = 0.0300 2 = 0.0600 3 = 0.1500 4 = 0.3600	1 = 0.0300 2 = 0.0600 3 = 0.1500 4 = 0.3600	0
SC-12	Remote access framework including policies and procedures established to provide secure access to telecommuting staff established for the management, monitoring, review, and audit of remote access to the organization.	1 = 0.60	1 = 0.05 2 = 0.10 3 = 0.25 4 = 0.60	Remote access to the SCADA system requires two-factor authentication.	0	1 = 0.0300 2 = 0.0600 3 = 0.1500 4 = 0.3600	0
SC-14	Network segregation. Firewalls, deep packet inspection, and/or application proxy gateways.	1 = 0.60	1 = 0.05 2 = 0.10 3 = 0.25 4 = 0.60	An actively managed firewall is in place to allow secure data transfer via DMZ.	1 = 0.0300 2 = 0.0600 3 = 0.1500 4 = 0.3600	1 = 0.0300 2 = 0.0600 3 = 0.1500 4 = 0.3600	0

(continued)

120

Table B.20 Manual vulnerability assessment table *(Continued)*

Control Code	Description	Priority	Status	Explanation/Example	Detect	Delay	Respond
SC-15	Logically separated control network. Minimal or single access points between corporate and control networks. Stateful firewall between corporate and control networks filtering on TCP and UDP ports. DMZ networks for data sharing.	1 = 0.60	1 = 0.05 2 = 0.10 3 = 0.25 4 = 0.60	An actively managed firewall is in place to allow secure data transfer via DMZ to provide operations data to utility asset managers.	1 = 0.0300 2 = 0.0600 3 = 0.1500 4 = 0.3600	1 = 0.0300 2 = 0.0600 3 = 0.1500 4 = 0.3600	0
SC-16	Defense-in-depth. Multiple layers of security with overlapping functionality.	1 = 0.60	1 = 0.05 2 = 0.10 3 = 0.25 4 = 0.60	A utility employs multiple types of physical and cybersecurity efforts to protect assets and systems. The efforts include things such as locking doors, physical access control, and unique login requirements for each staff member.	0	1 = 0.0300 2 = 0.0600 3 = 0.1500 4 = 0.3600	0
SC-17	Virtual Local Area Network (VLAN) for logical network segregation.	1 = 0.60	1 = 0.05 2 = 0.10 3 = 0.25 4 = 0.60	Within the SCADA system network, vendor systems are on a separate subnet.	0	1 = 0.0300 2 = 0.0600 3 = 0.1500 4 = 0.3600	0
SC-18	Minimize wireless network coverage.	1 = 0.60	1 = 0.05 2 = 0.10 3 = 0.25 4 = 0.60	Tests are conducted regularly to determine if the Wi-Fi signals reach outside the intended area of use. If the signal reaches outside the intended area, the signal is turned down accordingly.	0	1 = 0.0300 2 = 0.0600 3 = 0.1500 4 = 0.3600	0
SC-19	802.1X user authentication on wireless networks.	1 = 0.60	1 = 0.05 2 = 0.10 3 = 0.25 4 = 0.60	No "open" Wi-Fi connections are allowed.	0	1 = 0.0300 2 = 0.0600 3 = 0.1500 4 = 0.3600	0
SC-02	Centralized authentication system or single sign-on established to authorize access from a central system.	1 = 0.60	1 = 0.05 2 = 0.10 3 = 0.25 4 = 0.60	Operators have one username and password for PCS equipment, which is managed from a central system.	0	1 = 0.0300 2 = 0.0600 3 = 0.1500 4 = 0.3600	0
SC-20	Wireless equipment located on isolated network with minimal or single connection to control network.	1 = 0.60	1 = 0.05 2 = 0.10 3 = 0.25 4 = 0.60	Wi-Fi equipment in the plant does not connect directly to SCADA network.	0	1 = 0.0300 2 = 0.0600 3 = 0.1500 4 = 0.3600	0
SC-21	Unique wireless network identifier (SSID) for control network.	1 = 0.60	1 = 0.05 2 = 0.10 3 = 0.25 4 = 0.60	The Wi-Fi for the control system has a unique SSID from the business network.	0	1 = 0.0300 2 = 0.0600 3 = 0.1500 4 = 0.3600	0

(continued)

Table B.20 Manual vulnerability assessment table *(Continued)*

Control Code	Description	Priority	Status	Explanation/Example	Detect	Delay	Respond
SC-22	Separate Microsoft Windows domain for wireless (if using Windows).	1 = 0.60	1 = 0.05 2 = 0.10 3 = 0.25 4 = 0.60	A wireless LAN specific domain controller is in place.	0	1 = 0.0300 2 = 0.0600 3 = 0.1500 4 = 0.3600	0
SC-23	Wireless communications links encrypted.	1 = 0.60	1 = 0.05 2 = 0.10 3 = 0.25 4 = 0.60	All data transferred via the wireless network is encrypted using current wireless communication best practices.	0	1 = 0.0300 2 = 0.0600 3 = 0.1500 4 = 0.3600	0
SC-24	Communications links encrypted.	1 = 0.60	1 = 0.05 2 = 0.10 3 = 0.25 4 = 0.60	All data transferred via the wired network is encrypted using current wired communication best practices.	0	1 = 0.0300 2 = 0.0600 3 = 0.1500 4 = 0.3600	0
SC-25	Virtual Private Network (VPN) using IPsec, SSL, or SSH to encrypt communications from untrusted networks to the control system network.	1 = 0.60	1 = 0.05 2 = 0.10 3 = 0.25 4 = 0.60	An operator who can access the system remotely must do so through a secured VPN client configuration.	1 = 0.0300 2 = 0.0600 3 = 0.1500 4 = 0.3600	1 = 0.0300 2 = 0.0600 3 = 0.1500 4 = 0.3600	0
SC-03	Policies and procedures established for network segmentation including implementation of DMZs based on type and sensitivity of equipment, user roles, and types of systems established.	1 = 0.60	1 = 0.05 2 = 0.10 3 = 0.25 4 = 0.60	All external communication with the PCS is implemented via DMZ.	0	1 = 0.0300 2 = 0.0600 3 = 0.1500 4 = 0.3600	0
SI-01	Electronic commerce infrastructure in place providing integrity, confidentiality, and nonrepudiation and including adherence to pertinent laws, regulations, policies, procedures, and approval by management.	1 = 0.60	1 = 0.05 2 = 0.10 3 = 0.25 4 = 0.60	The company selected to perform billing is compliant with pertinent laws, regulations, policies, and procedures that are relevant to the utility.	0	1 = 0.0300 2 = 0.0600 3 = 0.1500 4 = 0.3600	0
SI-03	Interactive system for managing password implemented to ensure password strength.	1 = 0.60	1 = 0.05 2 = 0.10 3 = 0.25 4 = 0.60	When configuring a new user's password, it must meet minimum character length requirements.	0	1 = 0.0300 2 = 0.0600 3 = 0.1500 4 = 0.3600	0
AT-01	A general security awareness and response program established to ensure staff is aware of the indications of a potential incident, security policies, and incident response/ notification procedures.	2 = 0.25	1 = 0.05 2 = 0.10 3 = 0.25 4 = 0.60	An operator finds a USB media device. Based on their cybersecurity training, they know not to use it on the company network.	1 = 0.0125 2 = 0.0250 3 = 0.0625 4 = 0.1500	0	1 = 0.0125 2 = 0.0250 3 = 0.0625 4 = 0.1500

(continued)

Table B.20 Manual vulnerability assessment table *(Continued)*

Control Code	Description	Priority	Status	Explanation/Example	Detect	Delay	Respond
AT-02	Job-specific security training including incident response training for employees, contractors, and third-party users.	2 = 0.25	1 = 0.05 2 = 0.10 3 = 0.25 4 = 0.60	An operator has received what they believe to be a malicious email. They recognize that it is a phishing attack based on security training awareness programs the company has in place.	1 = 0.0125 2 = 0.0250 3 = 0.0625 4 = 0.1500	0	1 = 0.0125 2 = 0.0250 3 = 0.0625 4 = 0.1500
AU-05	Risk-based business continuity framework established under the auspices of the executive team to maintain continuity of operations and consistency of policies and plans throughout the organization. Another purpose of the framework is to ensure consistency across plans in terms of priorities, contact data, testing, and maintenance.	2 = 0.25	1 = 0.05 2 = 0.10 3 = 0.25 4 = 0.60	The facility has a documented and tested contingency plan to operate the facility without the use of SCADA software in the case of attack by ransomware.	0	0	1 = 0.0125 2 = 0.0250 3 = 0.0625 4 = 0.1500
AU-06	Policies and procedures established to validate, test, update, and audit the business continuity plan throughout the organization.	2 = 0.25	1 = 0.05 2 = 0.10 3 = 0.25 4 = 0.60	The business continuity plan is revised annually. Revisions are informed by planned exercises, actual events, or documented changes.	0	0	1 = 0.0125 2 = 0.0250 3 = 0.0625 4 = 0.1500
AU-07	Policies and procedures for system instantiation/deployment established to ensure business continuity.	2 = 0.25	1 = 0.05 2 = 0.10 3 = 0.25 4 = 0.60	The PCS has a testing/development environment to allow changes to be implemented without immediate effects on the production environment.	1 = 0.0125 2 = 0.0250 3 = 0.0625 4 = 0.1500	0	0
CM-03	Separation of duties implemented for user processes including risk of abuse.	2 = 0.25	1 = 0.05 2 = 0.10 3 = 0.25 4 = 0.60	Operators are only given clearance to areas they are expected to work in. Supervisors have the ability and training to monitor SCADA tech activities in the PCS.	1 = 0.0125 2 = 0.0250 3 = 0.0625 4 = 0.1500	1 = 0.0125 2 = 0.0250 3 = 0.0625 4 = 0.1500	0
CM-04	Separation of duties implemented for development, production, and testing work.	2 = 0.25	1 = 0.05 2 = 0.10 3 = 0.25 4 = 0.60	A SCADA technician must have a second technician review changes made to production equipment before they are implemented.	1 = 0.0125 2 = 0.0250 3 = 0.0625 4 = 0.1500	0	0

(continued)

Table B.20   Manual vulnerability assessment table *(Continued)*

Control Code	Description	Priority	Status	Explanation/Example	Detect	Delay	Respond
CM-05	SLAs for all third parties established, including levels of service and change controls.	2 = 0.25	1 = 0.05 2 = 0.10 3 = 0.25 4 = 0.60	A security policy that outlines what access permissions are distributed to third-party employees.	1 = 0.0125 2 = 0.0250 3 = 0.0625 4 = 0.1500	1 = 0.0125 2 = 0.0250 3 = 0.0625 4 = 0.1500	0
CM-07	Monitoring of resources and capabilities with notifications and alarms established to alert management when resources/ capabilities fall below a threshold.	2 = 0.25	1 = 0.05 2 = 0.10 3 = 0.25 4 = 0.60	IT monitors SCADA computers for processor usage that could indicate cryptojacking activity.	1 = 0.0125 2 = 0.0250 3 = 0.0625 4 = 0.1500	0	0
IR-01	Incident response program established with a formal Emergency Response Plan to restore systems and operations based on their criticality and within time constraints and effect recovery in case of a catalogue of disruptive events. Exercises conducted to test and revise plans and build organizational response capabilities.	2 = 0.25	1 = 0.05 2 = 0.10 3 = 0.25 4 = 0.60	Emergency Response Plan includes procedures for recovering SCADA system operation from system backup.	0	0	1 = 0.0125 2 = 0.0250 3 = 0.0625 4 = 0.1500
MA-03	Off-site equipment maintenance program including risk assessment of outside environmental conditions established.	2 = 0.25	1 = 0.05 2 = 0.10 3 = 0.25 4 = 0.60	The condition of off-site equipment and risk factors acting on the equipment are periodically reviewed and assessed via an independent party.	1 = 0.0125 2 = 0.0250 3 = 0.0625 4 = 0.1500	0	0
PM-03	Centralized logging system including policies and procedures to collect, analyze, and report to management.	2 = 0.25	1 = 0.05 2 = 0.10 3 = 0.25 4 = 0.60	A utility has a network intrusion detection system (NIDS) to monitor network traffic.	1 = 0.0125 2 = 0.0250 3 = 0.0625 4 = 0.1500	0	0
PM-04	SLAs for software and information exchange with internal/external parties in place including interfaces between systems and approved policies and procedures.	2 = 0.25	1 = 0.05 2 = 0.10 3 = 0.25 4 = 0.60	Third parties must review and sign an information-exchange policy before connecting to the system.	1 = 0.0125 2 = 0.0250 3 = 0.0625 4 = 0.1500	0	0
RA-01	Risk assessment and approval process before granting access to the organization's information systems.	2 = 0.25	1 = 0.05 2 = 0.10 3 = 0.25 4 = 0.60	A third-party system integrator would need to contact IT before connecting to the system's network.	1 = 0.0125 2 = 0.0250 3 = 0.0625 4 = 0.1500	0	0

(continued)

124

Table B.20 Manual vulnerability assessment table *(Continued)*

Control Code	Description	Priority	Status	Explanation/Example	Detect	Delay	Respond
RA-02	Third-party agreement process to ensure that external vendors and contractors utilize appropriate security measures for accessing, processing, communicating, or managing the organization's information or facilities.	2 = 0.25	1 = 0.05 2 = 0.10 3 = 0.25 4 = 0.60	System integrators can only access the facility's equipment remotely from a VPN connection.	1 = 0.0125 2 = 0.0250 3 = 0.0625 4 = 0.1500	1 = 0.0125 2 = 0.0250 3 = 0.0625 4 = 0.1500	0
SC-10	Program for hardening servers, workstations, routers, and other systems using levels of hardening based on criticality established. Program should include policies and procedures for whitelisting (deny-all, allow by exception).	2 = 0.25	1 = 0.05 2 = 0.10 3 = 0.25 4 = 0.60	Ports are disabled for all network devices when not in use.	1 = 0.0125 2 = 0.0250 3 = 0.0625 4 = 0.1500	1 = 0.0125 2 = 0.0250 3 = 0.0625 4 = 0.1500	0
SC-13	Testing standards including test data selection, protection, and system verification established to ensure system completeness.	2 = 0.25	1 = 0.05 2 = 0.10 3 = 0.25 4 = 0.60	Organization has a FAT procedure that requires vendors to demonstrate security of systems before they are purchased.	1 = 0.0125 2 = 0.0250 3 = 0.0625 4 = 0.1500	0	0
SC-04	Intrusion detection, prevention, and recovery systems including approved policies and procedures established to protect against cyberattacks. System includes repository of fault logging, analysis, and appropriate actions taken.	2 = 0.25	1 = 0.05 2 = 0.10 3 = 0.25 4 = 0.60	Monitoring of IDS is conducted to determine if ongoing attacks are occurring and incidence response actions have been documented.	1 = 0.0125 2 = 0.0250 3 = 0.0625 4 = 0.1500	1 = 0.0125 2 = 0.0250 3 = 0.0625 4 = 0.1500	1 = 0.0125 2 = 0.0250 3 = 0.0625 4 = 0.1500
SC-05	Anomaly-based IDS/ IPS established including policies and procedures.	2 = 0.25	1 = 0.05 2 = 0.10 3 = 0.25 4 = 0.60	The IT tech monitors IDS system exception logs daily to determine if ongoing attacks are occurring and works with SCADA tech to address any issues.	1 = 0.0125 2 = 0.0250 3 = 0.0625 4 = 0.1500	0	1 = 0.0125 2 = 0.0250 3 = 0.0625 4 = 0.1500
SC-06	Network management and monitoring established including deep packet inspection of traffic, QoS, port-level security, and approved policies and procedures.	2 = 0.25	1 = 0.05 2 = 0.10 3 = 0.25 4 = 0.60	An actively managed firewall is in place to allow secure data transfer via DMZ to provide operations data to utility asset managers.	1 = 0.0125 2 = 0.0250 3 = 0.0625 4 = 0.1500	1 = 0.0125 2 = 0.0250 3 = 0.0625 4 = 0.1500	0

(continued)

Table B.20 Manual vulnerability assessment table *(Continued)*

Control Code	Description	Priority	Status	Explanation/Example	Detect	Delay	Respond
SC-07	Information-exchange protection program in place to protect data in transit through any communication system including the Internet, email, and text messaging and approved policies and procedures.	2 = 0.25	1 = 0.05 2 = 0.10 3 = 0.25 4 = 0.60	Web applications for SCADA software use encryption to protect data in transit.	0	1 = 0.0125 2 = 0.0250 3 = 0.0625 4 = 0.1500	0
SC-08	Routing controls established to provide logical separation of sensitive systems and enforce the organization's access control policy.	2 = 0.25	1 = 0.05 2 = 0.10 3 = 0.25 4 = 0.60	Within the SCADA system network, vendor systems are placed on a separate subnet rather than being on a single "flat" network.	0	1 = 0.0125 2 = 0.0250 3 = 0.0625 4 = 0.1500	0
SC-09	Process isolation established to provide a manual override "air gap" between highly sensitive systems and regular environments.	2 = 0.25	1 = 0.05 2 = 0.10 3 = 0.25 4 = 0.60	A manual method for disconnecting the PCS network from other networks is implemented and documented.	0	0	1 = 0.0125 2 = 0.0250 3 = 0.0625 4 = 0.1500
SI-02	System acceptance standards including data validation (input/output), message authenticity, and system integrity established to detect information corruption during processing.	2 = 0.25	1 = 0.05 2 = 0.10 3 = 0.25 4 = 0.60	Acquired assets are inspected, assessed, and documented before implementation with existing systems.	1 = 0.0125 2 = 0.0250 3 = 0.0625 4 = 0.1500	0	0
SI-05	Privileged programs controls established to restrict usage of utility programs that could reset passwords or override controls as well as IT audit tools that can modify or delete audit data.	2 = 0.25	1 = 0.05 2 = 0.10 3 = 0.25 4 = 0.60	Utility has implemented tiered access so non-administrator users are unable to make changes to system security settings.	0	1 = 0.0125 2 = 0.0250 3 = 0.0625 4 = 0.1500	0
AU-08	Template for the organization's confidentiality/nondisclosure agreements defined, reviewed, and approved periodically by management.	3 = 0.10	1 = 0.05 2 = 0.10 3 = 0.25 4 = 0.60	Reviews of the organization's confidentiality/nondisclosure agreements are periodically scheduled by a responsible party.	1 = 0.0050 2 = 0.0010 3 = 0.0250 4 = 0.0600	1 = 0.0050 2 = 0.0010 3 = 0.0250 4 = 0.0600	0

(continued)

Table B.20 Manual vulnerability assessment table *(Continued)*

Control Code	Description	Priority	Status	Explanation/Example	Detect	Delay	Respond
CIE-01	A program is in place to engage engineering staff in understanding and mitigating high-consequence and constantly evolving cyber threat throughout the engineering life cycle, including design, implementation, maintenance, and decommissioning.	3 = 0.10	1 = 0.05 2 = 0.10 3 = 0.25 4 = 0.60	Engineering staff is fully aware of the potential for a cyber breach. They design electrical and mechanical systems to provide functionality in the case of a SCADA system compromise.	1 = 0.0050 2 = 0.0010 3 = 0.0250 4 = 0.0600	1 = 0.0050 2 = 0.0010 3 = 0.0250 4 = 0.0600	0
CM-01	Policies for defining business requirements including data validation and message authenticity established to ensure that new/upgraded systems contain appropriate security requirements and controls.	3 = 0.10	1 = 0.05 2 = 0.10 3 = 0.25 4 = 0.60	Policies to define minimum security features (e.g., secure protocols, active directory integration) required for new systems. This could include review and approval by change management and/or security team.	1 = 0.0050 2 = 0.0010 3 = 0.0250 4 = 0.0600	1 = 0.0050 2 = 0.0010 3 = 0.0250 4 = 0.0600	0
CM-02	Procedure modification tracking program in place to manage and log changes to policies and procedures.	3 = 0.10	1 = 0.05 2 = 0.10 3 = 0.25 4 = 0.60	The Emergency Response Plan is stored in a central repository and clearly displays the version and date of when it was implemented.	0	0	1 = 0.0050 2 = 0.0010 3 = 0.0250 4 = 0.0600
IA-08	Policies for security of stand-alone, lost, and misplaced equipment in place.	3 = 0.10	1 = 0.05 2 = 0.10 3 = 0.25 4 = 0.60	An operator misplaces a managed phone. Based on the missing-equipment policy, they contact IT to report the device lost.	0	0	1 = 0.0050 2 = 0.0010 3 = 0.0250 4 = 0.0600
IR-03	A legal/contractual/regulatory framework established with a formal Emergency Response Plan to track legal/contractual/regulatory requirements and the efforts to meet them with respect to each important system within the organization. Another purpose of the framework is to ensure compliance of policies and procedures with privacy laws, handling cryptographic products, intellectual property rights, and data-retention requirements.	3 = 0.10	1 = 0.05 2 = 0.10 3 = 0.25 4 = 0.60	The Emergency Response Plan is reviewed and updated once a year by responsible staff.	0	0	1 = 0.0050 2 = 0.0010 3 = 0.0250 4 = 0.0600

(continued)

127

Table B.20   Manual vulnerability assessment table *(Continued)*

Control Code	Description	Priority	Status	Explanation/Example	Detect	Delay	Respond
MP-02	Information exit mechanisms in place to prevent data and/or software leaving premises without authorization or logging.	3 = 0.10	1 = 0.05 2 = 0.10 3 = 0.25 4 = 0.60	An approved data leakage prevention (DLP) system is implemented or manual procedures to control data and/or software leaving organization.	1 = 0.0050 2 = 0.0010 3 = 0.0250 4 = 0.0600	1 = 0.0050 2 = 0.0010 3 = 0.0250 4 = 0.0600	0
PM-01	An asset inventory of all electronic components including model, software/firmware, etc., that is maintained and referenced when vendor vulnerabilities are disclosed.	3 = 0.10	1 = 0.05 2 = 0.10 3 = 0.25 4 = 0.60	A database is used to keep track of building conditions in the facility.	1 = 0.0050 2 = 0.0010 3 = 0.0250 4 = 0.0600	0	0
PM-02	Policies and procedures for acceptable use of assets and information approved and implemented.	3 = 0.10	1 = 0.05 2 = 0.10 3 = 0.25 4 = 0.60	PLCs that cannot update past a specific security revision are not acceptable for use in the PCS.	0	1 = 0.0050 2 = 0.0010 3 = 0.0250 4 = 0.0600	0
PS-01	Policies and procedures for hiring/terminating processes on employees, contractors, or support companies to include background checks and contract agreements approved and implemented.	3 = 0.10	1 = 0.05 2 = 0.10 3 = 0.25 4 = 0.60	A background check on employees is required before they may be given access to the PCS system.	1 = 0.0050 2 = 0.0010 3 = 0.0250 4 = 0.0600	1 = 0.0050 2 = 0.0010 3 = 0.0250 4 = 0.0600	0
PS-02	Defined and approved security roles and responsibilities of all employees, contractors, and third-party users.	3 = 0.10	1 = 0.05 2 = 0.10 3 = 0.25 4 = 0.60	A company policy is in place limiting the access of third-party users to assets, systems, and data.	1 = 0.0050 2 = 0.0010 3 = 0.0250 4 = 0.0600	1 = 0.0050 2 = 0.0010 3 = 0.0250 4 = 0.0600	1 = 0.0050 2 = 0.0010 3 = 0.0250 4 = 0.0600
PS-03	A clear desk policy in place including clear papers, media, desktop, and computer screens.	3 = 0.10	1 = 0.05 2 = 0.10 3 = 0.25 4 = 0.60	Confidential documents are stored in locked file cabinets when not in use, as required by policy.	0	1 = 0.0050 2 = 0.0010 3 = 0.0250 4 = 0.0600	0
PS-04	Disciplinary process for security violations established.	3 = 0.10	1 = 0.05 2 = 0.10 3 = 0.25 4 = 0.60	An operator who props open doors to critical areas could face disciplinary action as outlined in the utility's policies and procedures.	0	1 = 0.0050 2 = 0.0010 3 = 0.0250 4 = 0.0600	0
SA-01	Authorization process established for new systems or changes to existing information processing systems.	3 = 0.10	1 = 0.05 2 = 0.10 3 = 0.25 4 = 0.60	A change management/review process is used to evaluate suggested changes to facility.	0	1 = 0.0050 2 = 0.0010 3 = 0.0250 4 = 0.0600	0

(continued)

Table B.20 Manual vulnerability assessment table *(Continued)*

Control Code	Description	Priority	Status	Explanation/Example	Detect	Delay	Respond
SA-02	Change controls of systems development, outsourced development, system modification, and testing established, including acceptance criteria for new systems, monitoring of internal/outsourced development, and control of system upgrades.	3 = 0.10	1 = 0.05 2 = 0.10 3 = 0.25 4 = 0.60	A third-party system integrator is preparing to make changes to SCADA software. The SCADA tech requires the integrator to follow the change procedure and test the changes in a sandbox environment before they are deployed in production.	0	1 = 0.0050 2 = 0.0010 3 = 0.0250 4 = 0.0600	0
SA-03	Change controls of operating systems, network configuration/topology, and network security established, including changes to IDS/IPS, traffic control/monitoring, new systems, and system upgrades.	3 = 0.10	1 = 0.05 2 = 0.10 3 = 0.25 4 = 0.60	Automatic updates to the operating system are disabled, but monthly manual updates are reviewed and applied in coordination with operations.	0	1 = 0.0050 2 = 0.0010 3 = 0.0250 4 = 0.0600	0
SA-04	Risk-based mobility policies and procedures established to protect against inherent risk of mobile computing and communication systems.	3 = 0.10	1 = 0.05 2 = 0.10 3 = 0.25 4 = 0.60	Remote access is restricted to only the most necessary applications and only allowed through secure measures.	1 = 0.0050 2 = 0.0010 3 = 0.0250 4 = 0.0600	1 = 0.0050 2 = 0.0010 3 = 0.0250 4 = 0.0600	0
SA-05	Periodic review of backup policies and procedures and testing of recovery processes.	3 = 0.10	1 = 0.05 2 = 0.10 3 = 0.25 4 = 0.60	System backups are tested on a regular basis by completing a system restoration to the test environment.	0	0	1 = 0.0050 2 = 0.0010 3 = 0.0250 4 = 0.0600
SI-04	Organization-wide clock synchronization system in place.	3 = 0.10	1 = 0.05 2 = 0.10 3 = 0.25 4 = 0.60	All managed network devices synchronize their clocks to a known good source.	0	1 = 0.0050 2 = 0.0010 3 = 0.0250 4 = 0.0600	0
CM-06	Risk-based policies and procedures for change controls, reviews, and audits of SLAs.	4 = 0.05	1 = 0.05 2 = 0.10 3 = 0.25 4 = 0.60	Inviting all affected parties to discussions to prevent the development of vulnerabilities in the facility.	1 = 0.0025 2 = 0.0050 3 = 0.0125 4 = 0.0300	0	0
IA-02	Access control for the management, monitoring, review, and audit of accounts established including access control, account roles, privilege accounts, password policies, and executive oversight.	4 = 0.05	1 = 0.05 2 = 0.10 3 = 0.25 4 = 0.60	Upon staff termination or resignation, login credentials are disabled as part of the Human Resources process.	0	1 = 0.0025 2 = 0.0050 3 = 0.0125 4 = 0.0300	0

(continued)

Table B.20 Manual vulnerability assessment table *(Continued)*

Control Code	Description	Priority	Status	Explanation/Example	Detect	Delay	Respond
IR-02	A security program established with a formal Emergency Response Plan to respond to security incidents; monitor, discover, and handle security alerts and technical vulnerabilities; collect and analyze security data; limit the organization's risk profile; and ensure that management is aware of changing/emerging risks.	4 = 0.05	1 = 0.05 2 = 0.10 3 = 0.25 4 = 0.60	A SCADA tech believes a machine is infected and responds according to the utility's emergency response plan for cybersecurity-based incidents.	0	0	1 = 0.0025 2 = 0.0050 3 = 0.0125 4 = 0.0300
MA-01	A controlled maintenance system is in place to organize, schedule, document, and monitor the maintenance and repairs performed on information system assets in the PCS.	4 = 0.05	1 = 0.05 2 = 0.10 3 = 0.25 4 = 0.60	Based on the company's controlled maintenance program, a utility will format network devices to factory settings before sending them out of the organization for maintenance.	1 = 0.0025 2 = 0.0050 3 = 0.0125 4 = 0.0300	1 = 0.0025 2 = 0.0050 3 = 0.0125 4 = 0.0300	0
MA-02	Maintenance of relationships with authorities, professional associations, interest groups, etc., formalized. This is done, in part, to maintain an up-to-date situational awareness of relevant threats.	4 = 0.05	1 = 0.05 2 = 0.10 3 = 0.25 4 = 0.60	The utility is a member of DHS's CISA mailing list to receive frequent communications on PCS vulnerabilities discovered and patches available. SCADA techs regularly review alerts to determine if the alerts are applicable to their system.	1 = 0.0025 2 = 0.0050 3 = 0.0125 4 = 0.0300	0	1 = 0.0025 2 = 0.0050 3 = 0.0125 4 = 0.0300
MP-01	Storage media management and disposal program established to ensure that any sensitive data/software is used appropriately and is removed prior to media disposal (including approved policies and procedures).	4 = 0.05	1 = 0.05 2 = 0.10 3 = 0.25 4 = 0.60	When decommissioning a network device that was used in the production environment, IT is required to return it to factory conditions before it leaves the facility.	1 = 0.0025 2 = 0.0050 3 = 0.0125 4 = 0.0300	0	1 = 0.0025 2 = 0.0050 3 = 0.0125 4 = 0.0300
MP-03	Policies and procedure repository in place to be available to all authorized staff.	4 = 0.05	1 = 0.05 2 = 0.10 3 = 0.25 4 = 0.60	Company policies and procedures are available in a central, secure, shared location.	1 = 0.0025 2 = 0.0050 3 = 0.0125 4 = 0.0300	0	1 = 0.0025 2 = 0.0050 3 = 0.0125 4 = 0.0300
PM-05	Data classification policies and procedures for handling and labeling based on confidentiality and criticality approved and implemented.	4 = 0.05	1 = 0.05 2 = 0.10 3 = 0.25 4 = 0.60	A policy to store and manage access to PLC programs.	1 = 0.0025 2 = 0.0050 3 = 0.0125 4 = 0.0300	0	0

(continued)

Table B.20 Manual vulnerability assessment table *(Continued)*

Control Code	Description	Priority	Status	Explanation/Example	Detect	Delay	Respond
SU-01	A supply chain risk management program.	3 = 0.10	1 = 0.05 2 = 0.10 3 = 0.25 4 = 0.60	Chain-of-custody documentation is required for all chemicals used in treatment.	1 = 0.0050 2 = 0.0100 3 = 0.0250 4 = 0.0600	0	0
SU-02	A supply chain risk management program that includes cybersecurity.	4 = 0.05	1 = 0.05 2 = 0.10 3 = 0.25 4 = 0.60	Preferred vendors for computer hardware, software, and peripherals are identified and selected based on evaluation of their supply chain among other criteria.	1 = 0.0025 2 = 0.0050 3 = 0.0125 4 = 0.0300	0	0

SECTION B.8:   REFERENCES

The following references were used throughout the document. Please research if more recent information is available.

AWWA CRAT:

American Water Works Association—Cybersecurity Risk Assessment Tool

Cybersecurity & Guidance | American Water Works Association (awwa.org)

https://www.awwa.org/Resources-Tools/Resource-Topics/Risk-Resilience/Cybersecurity-Guidance

CISA:

Cybersecurity & Infrastructure Agency

Cost of a Cyber Incident: Systematic Review and Cross-Validation | CISA

https://www.cisa.gov/publication/cost-cyber-incident-systematic-review-and-cross-validation

IBM Breach Report:

International Business Machines

Cost of a Data Breach Report 2021 | IBM

https://www.ibm.com/security/data-breach

USEPA WHEAT:

US Environmental Protection Agency Water Health and Economic Analysis Tool

Download and Install the Water Health and Economic Analysis Tool for Your Water Utility | US EPA

https://www.epa.gov/waterriskassessment/download-and-install-water-health-and-economic-analysis-tool-your-water-utility.

The web version of WHEAT is contained within EPA VSAT. https://vsat.epa.gov/vsat/

APPENDIX C

Natural Hazards Introduction

SECTION C.1:   NATURAL HAZARDS OVERVIEW

Sec. C.1.1   Introduction

A risk analysis consistent with J100-21 is incomplete unless natural hazards are considered. The natural hazards of concern within J100-21 are seismic activity, hurricanes, tornadoes, wildfires, ice storms, floods, and any hazard that is locally significant, such as extreme temperatures, avalanche, tsunami, landslides, and mudslides. The risk of these events for any given asset is analyzed by estimating its frequency (usually based on historic frequencies) and the effects on the asset (based on its type of structure, age, condition, and other factors).

The severity and frequency of natural hazards generally depend on the geographical location of the facility or asset. For example, seismic activity is much more likely to occur on the West Coast of the United States, Hawaii, Alaska, and the US Virgin Islands, whereas hurricane risk is greater along the East and Gulf Coasts and Florida. When considering natural hazards, the magnitude and expected frequency of the event are typically determined from historical data. The geographical distribution and frequency of occurrence data of these initiating events can be obtained from various governmental agencies. Source information will be discussed in the following sections.

Unlike terrorism events, consideration of a facility's ability to withstand most natural hazards of specified intensity is included in building and structure design codes. In almost all areas of the United States, the local, state, or national statutes require that new construction meets the structural requirements of the Uniform Building Code (UBC) or, more recently, the International Building Code (IBC). Every municipality or county typically has a building department that performs a plan check for new construction and revisions to existing structures. Once the plan check is approved, a building permit will

133

be issued, and on-site inspections are conducted at key points in the construction process. In cases not covered by local statutes, the financial institution providing the loan or the insurance carrier generally requires that the building be designed and constructed in accordance with the IBC. For the purposes of J100-21, all hazard magnitudes equal to or less than the code requirements at the time of construction or major renovation are assumed to have no major damage or loss of service beyond routine maintenance.

The Federal Emergency Management Agency (FEMA) maintains The National Risk Index* that can be referenced when identifying risks in a specific location. It is updated regularly to ensure historical information is available and can be filtered so that the user can pinpoint specific areas of interest.

The next six appendixes describe the process of evaluating the risks of natural hazards. First there will be an introduction to risk and the concepts of consequences, vulnerability, and likelihood/frequency followed by specific guidance for each of the six types of natural hazards listed here. The details for each natural hazard provide references for how to determine consequences, vulnerability, and likelihood/frequency, as well as examples.

- Appendix D Flood
- Appendix E Hurricane
- Appendix F Ice Storm
- Appendix G Seismic
- Appendix H Tornado
- Appendix I Wildfire

Sec. C.1.2 Estimate Risk

To estimate a system's risk from a specific hazard, the hazard frequency/return period, vulnerability, and consequence of failure are calculated for each asset. Once this is established, the overall system risk is determined by combining risks across multiple scenarios. The risk of each asset can be calculated using the standard risk equation:

$$R_i = C_i \times V_i \times T_i$$

* https://hazards.fema.gov/nri/

134

Where:

R_i = Risk (measured in deaths, severe injuries, or dollar losses) of an initiating event i.

C_i = Consequences of the considered hazard-initiating event under the "worst reasonable case" assumption. This should include the J100-21 consequence categories that encompass casualties (fatalities and severe injuries), unmet demand, dollar losses to the facility's owner, and/or dollar losses to the regional economy. Losses to the owner would consist of repair and/or replacement cost depending on the severity of damage; lost net revenue due to down time, liability costs, etc.; and other direct losses because of damage to the asset by the initiating event i. Losses for damage to individual facilities and system outage losses should be included.

V_i = Vulnerability of the structure or equipment for the type of event considered. For example, the vulnerability of underground piping to a hurricane would be very low. However, the vulnerability of this same pipeline to earthquake could be very high. Modern cast-in-place concrete structures should be resistant to both wind and earthquake loading. Vulnerability is expressed as the likelihood of an event having the consequences described as the worst reasonable case, given that the event occurs.

T_i = Threat likelihood of an initiating event i. The frequency of an event is usually inversely correlated with the intensity. The likelihood/frequency is calculated on an annual basis.

The subscript "i" indicates that losses are a function of the hazard considered. The risk contribution is a function of its annualized contribution, so frequently occurring events have more weight.

The next three sections contain additional details regarding application of this conceptual approach for each of the six types of natural hazards previously described.

135

Sec. C.1.3 General—Frequency of the Initiating Event and Associated Hazard Intensity

Natural hazard intensity can be defined by categories of severity and likelihood. This may be in terms of ground-shaking intensity for earthquakes, depth of inundation and water velocity for floods, and wind speed for hurricanes and tornadoes. The J100-21 Reference Threats show the levels of natural hazards commonly used in the sector. Generally, as the expected intensity of a natural hazard event increases, the expected frequency decreases. The frequency is often expressed in terms of return period, which can also be related to the probability of occurrence in a set time period. For example, an earthquake ground motion with a 475-year return period (or nominally 500-year) can be defined for a site. This 475-year event has a 10 percent probability of being exceeded in a 50-year period. Further, floodplain limits are also often defined with a 500-year return period, similarly having a 10 percent probability of being exceeded in a 50-year period.

For many years, structural seismic design codes and standards set the earthquake return period and intensity to be considered in design. In addition, inherent in the code requirements were performance expectations regarding collapse avoidance under intense shaking. Current codes are based on consideration of the probabilistic seismic hazard at each site and a probabilistic estimate of the margin against collapse inherent in structures designed to the seismic provisions in the standard (e.g., collapse fragility). That is, ground motions take into account earthquakes over a range of sizes and from multiple source zones; code-based ground motions do not represent a single earthquake. For system analyses, use of deterministic scenarios representing a range of return periods and sources (for earthquakes) is preferred rather than using probabilistic hazards and associated intensities.

Some natural hazards can also result in secondary hazards; for example, a secondary earthquake hazard might be permanent ground deformation that results from shaking, liquefaction, landslides, and surface fault rupture. Secondary hazards from hurricanes can include tornadoes, storm surges, and tsunamis. Secondary hazards may be very damaging and may need to be defined and analyzed in the assessment as

136

a separate primary event, if important to the region. Risks for primary events can be combined.

Sec. C.1.4 General—Asset Vulnerability

Assessing the vulnerability of an asset to the natural hazard is a fundamental part of risk assessment. Vulnerability of a particular asset is often determined by selecting a fragility curve for a particular hazard parameter and asset type (see NIST Guide Brief No. 4[†]).

A fragility curve relates hazard intensity to an expected level or extent of damage. As the hazard intensity increases, the level of damage also increases. The level of damage can be expressed in terms of percent of replacement cost, percent operability following the event, expected outage/repair time, and (for pipelines) the number of failures per unit length.

For natural hazards such as earthquake, flood, and wind, fragility relationships for many infrastructure components can be found in Hazards US (HAZUS). HAZUS is a natural hazard loss-estimation methodology developed by the National Institute of Building Standards. It includes a library of fragilities for buildings and structures, including tanks, over a range of intensities. The fragility library was developed based on empirical data, analytical methods, and expert opinion. Other sources of fragility relationships are available. Further, for important structures, the owner may want to develop facility-specific fragility relationships using analytical methods.

Selection and modification of an existing fragility relationship are an important part of establishing vulnerability of an asset or system. Selection and development of fragility relationships should consider the elements appropriate to the hazard.

Sec. C.1.5 General—Consequences of Failure

Partial or complete failure of an asset will have some consequence. Loss of functionality and associated impact on the overall system may be significant. Other consequences can include inundation of a downstream population if a dam or reservoir fails. In this scenario, the consequence of failure would necessarily include cost of repairs of the

---

[†] http://nvlpubs.nist.gov/nistpubs/SpecialPublications/NIST.SP.1190GB-4.pdf

asset as well as costs associated with damage resulting from the failed asset.

Cascading impact (how the failure of one system can impact another sector) needs to be considered as well. For instance, the loss of potable water for a hospital may have an impact on the healthcare sector, as patients scheduled for surgery may need to be relocated to a hospital with system water. This in turn may overload ambulatory services, which in turn cannot support other requests. The cascading impact needs to be mapped out and can involve multiple sectors.

Water and wastewater system risk can be assessed by considering the impact of expected damage to all system components when subjected to a given scenario. It is recommended that a series of scenarios representing a range of return periods and hazard sources be taken into consideration for the risk assessment. Depending upon the desired outcome of the risk assessment, this task can be performed using various levels of sophistication.

The simplest approach to risk assessment is to use a workshop approach. This assessment engages personnel familiar with the operation of the system under study. The expected damage to key system components is presented using GIS and system schematics to show the significance of each asset within the system. Typically, consensus can be developed on the likely performance of the system for predetermined scenarios.

If a more sophisticated approach is desired, quantification of results using a numerical model of the system can be developed considering the hazards and asset vulnerabilities. Various attempts have been made in the academic community to develop probabilistic-based software to perform hydraulic network analyses of damaged systems; however, none is commercially available.

From a risk perspective, system outage (people-outage days) is generally the largest economic contributor to the loss. However, direct damage (the cost to repair damaged assets) should also be considered.

APPENDIX D

Flood

SECTION D.1:   INTRODUCTION

Floods are a common hazard in the United States and can range dramatically in their area of impact. Some flood events develop slowly over time, while others occur quickly without warning. Flooding can result from events such as thunderstorms, hurricanes, winter storms, snowmelt, and dam releases or failures. Inundation events are typically described in terms of their statistical frequency: F1 (100-year flood) and F2 (500-year flood). These data are used to demarcate floodplain boundaries and determine high-risk areas for mitigation planning. For the purposes of J100, storm surge due to hurricanes is calculated under the hurricane hazard (Appendix E).

Flood events typically inflict water damage to equipment and building contents rather than major structural damage that compromises the integrity of a facility. Water damage to drywall, electrical fixtures, etc., and mold damage may render a flooded structure unusable or so costly to restore that it is considered a total loss. Assets subjected to flood hazards may be sorted into two categories: protected and unprotected. Protected facilities include some sort of mitigation measure in place to prevent inundation by floodwaters (dikes, sandbags, elevated structure, special construction, etc.). For these types of assets, there will be no damage until water levels exceed the protection element, at which point the asset would begin to experience flood damage. Unprotected assets will suffer increased damage and recovery time as water depth increases to a predefined level (100% loss depth). In many cases, the operation of an asset may be terminated based on owner/operator decisions at a predetermined threshold floodwater elevation.

To determine whether an asset is subject to flooding, reference FEMA Flood Insurance Rate Maps (FIRMs) which are available at https://msc.fema.gov/portal. An example FEMA FIRM is provided in Figure D.1.

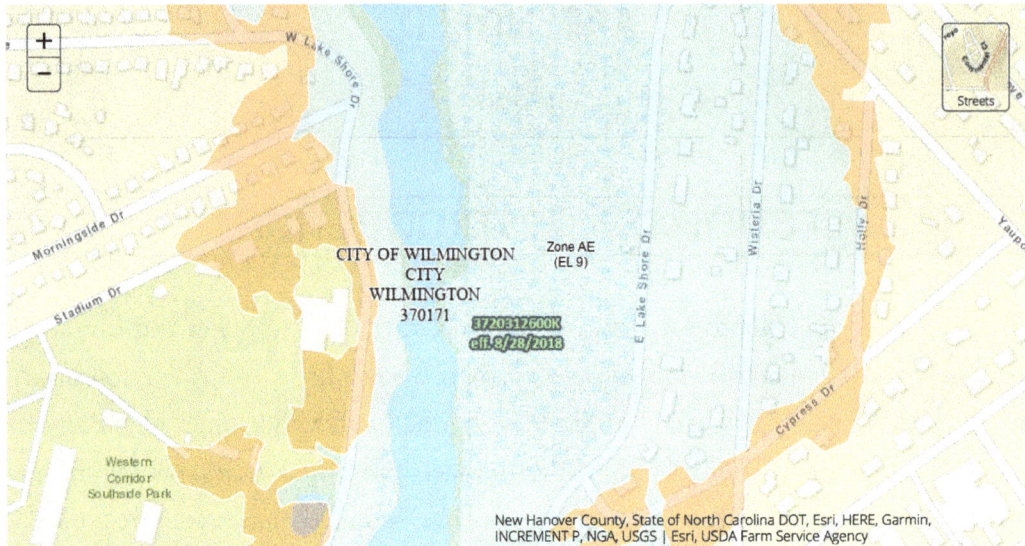

Figure D.1   Example FEMA Flood Insurance Map (FIRM)

The 1% annual flood (100-year flood), also known as the base flood, is the flood that has a 1% annual chance of being equaled or exceeded in any given year. The Base Flood Elevation (BFE) is the elevation to which floodwater is anticipated to rise during the base flood.

Flood hazard areas on the FIRM are identified as a Special Flood Hazard Area (SFHA), which is defined as the area that will be inundated by the flood event having a 1% chance of being equaled or exceeded in a given year (i.e., 100-year flood). SFHAs are labeled as Zone A, Zone AO, Zone AH, Zones A1–A30, Zone AE, Zone A99, Zone AR, Zone AR/AE, Zone AR/AO, Zone AR/A1–A30, Zone AR/A, Zone V, Zone VE, and Zones V1–V30.

Moderate flood hazard areas, labeled Zone B or Zone X (shaded), are also shown on the FIRMs and represent the area between the limits of the base flood and the 0.2% annual chance (or 500-year) flood.

Zones outside of the 0.2% chance flood are labeled Zone C or Zone X (unshaded).

FEMA flood zone guidance is as follows:

Moderate- to Low-Risk Areas
Zones B, C, and X. Assume average flood depths are less than 1 ft.

High-Risk Areas

Zone A.   Areas subject to inundation by the 1% annual chance flood event generally determined using approximate methodologies. Because detailed hydraulic analyses have not been performed, no BFEs or flood depths are shown.

Zone AE and A1–A30.   Areas subject to inundation by the 1% annual chance flood event determined by detailed methods. BFEs are shown.

Zone AH.   Areas subject to inundation by 1% annual chance shallow flooding (usually areas of ponding) where average depths are 1–3 ft. BFEs derived from detailed hydraulic analyses are shown in this zone.

Zone AO.   Areas subject to inundation by 1% annual chance shallow flooding (usually sheet flow on sloping terrain) where average depths are 1–3 ft.

Zone AR.   Areas that result from the decertification of a previously accredited flood protection system that is determined to be in the process of being restored to provide base flood protection.

Flood Zone AR/AE.   Flood zone AR/AE areas have a 1 percent chance of annual flooding as determined by FEMA's BFE, which is the regulatory requirement for the elevation floodproofing structures. Areas in this designation contain a mix of flood zone AR and flood zone AE in the same area.

Flood Zone AR/AO.   Flood zone AR/AO is an area with a combination of flood zones AR and AO and is given at least a 1 percent chance of annual flooding. In these areas, some homes and businesses may be protected by a water control system, such as a dam, while others are not.

Flood Zone AR/A1–A30.   Flood zone AR/A1–A30 holds a particularly distinct designation due to the temporary risk of increased flood damage given a flood control system in the area may be undergoing restoration or reconstruction. Some homes and businesses in these areas may be much more prone to flooding than others based on the location of the flood control system.

Flood Zone AR/A.   Flood zone AR/A is another zone designated by FEMA which contains a flood control system that has been temporarily decommissioned or is undergoing repairs or restoration. These areas are especially risky because the weakened flood control system is coupled with the 1 percent chance of annual flooding and a 25 percent chance of flooding at least once during a 30-year mortgage.

Zone A99. Areas subject to inundation by the 1% annual chance flood event but which will ultimately be protected upon completion of an under-construction federal flood protection system. No BFEs or depths are shown.

High Risk—Coastal Areas and Zone V

Zone V. Areas along coasts subject to inundation by the 1% annual chance flood event with additional hazards associated with storm-induced waves.

Zone VE and V1–30. Areas subject to inundation by the 1% annual chance flood event with additional hazards due to storm-induced velocity wave action. BFEs derived from detailed hydraulic analyses are shown.

Undetermined Risk Area

Zone D. Areas with possible but undetermined flood hazards. Use best judgment on a case-by-case basis.

SECTION D.2: CONSEQUENCES

Financial losses to the owner are the sum of repair and/or replacement costs, net revenue lost due to the incident, and other costs depending on the incident (e.g., debris removal, casualty liabilities). If the structure is fairly impervious to water damage, repair costs could be minimal, but there still may be significant revenue losses if the asset cannot function during the time of flooding and cleanup. In a single facility with multiple assets located at different elevations, each asset should be assessed at the height of the water during the flood being analyzed. Additional consequences can include serious injuries, fatalities, and/or economic loss to the community.

The process for calculating financial loss due to flood includes five steps and should be completed for both the 1% (100-year flood) and 0.2% (500-year flood) annual chance flood event as applicable:

1. Locate the asset on the FEMA FIRM (https://msc.fema.gov/portal). Continue to the second step if the asset lies within one of the applicable flood zones. If the asset lies in an unshaded zone (usually marked as C or X), there is no risk to the defined level of flooding.

2. Estimate the water depth for the assessed flood. Depending on the flood zone, a BFE may be provided, which provides the elevation to which floodwater is anticipated to rise during the base flood. This elevation can be compared to the asset's grade elevation to determine the expected water depth at the asset. If the asset grade elevation is not known, it can be estimated from a topographical map (https://apps.nationalmap.gov/downloader/#/). If BFE is not provided on the FIRM, other methods/sources can be used to estimate water depth, such as historical USGS river gauge data or historical records from the facility or locality.

3. Estimate the cost to repair/replace the asset and/or its damaged components, including cleanup costs. Determine which asset components will be damaged or ruined by standing water and their replacement costs. If the building is constructed of water-tolerant materials, then much less damage is expected than for materials that are ruined when water-soaked. Similarly, if electrical components are subject to inundation, such as in underground conduit, manholes, and trenches, and are not waterproof, then it must be assumed there will be extensive damage. Mechanical equipment such as piping, pumps, valves, and tanks may not be damaged, but the controls, motors, electrical and communication equipment, thermocouples, etc., may need replacement or repair. Also consider cleanup associated with residual mud and debris, mold, rot, and damage to carpets, drapes, furniture, and equipment that is sensitive to oxidation (rusting).

4. Determine if the damage will result in service denial; if so, estimate the downtime required to repair or replace the assets. Factor in redundancy within the system as well as contingency plans when estimating downtime. Calculate the lost revenue associated with the service denial.

5. The total loss will consist of the sum of the repair and replacement costs plus the cleanup cost and lost revenue due to service denial.

Consequences of flooding may also include fatalities and serious injuries, which should be estimated based on the amount of warning the asset's personnel would have been given for the flooding event.

SECTION D.3:   VULNERABILITY

The vulnerability of an asset to flooding depends on many factors. The first factor is the asset location. If the asset is not in the floodplain being assessed, then the asset does not carry a risk to that level flood. If the asset is in the floodplain, then the water depth and potential to damage the structure or other components will need to be considered. To evaluate the vulnerability of an asset to the 100-year and 500-year floods, the below assessment can be completed. A separate assessment will need to be done for each flood as anticipated water levels will differ.

Pre-Question: Is the asset at least partially located in the flood zone being assessed? If not, the Vulnerability defaults to 0. If so, move on to the questions below. Flood zone information can be obtained from a FIRM, which can be accessed at http://msc.fema.gov.

1. Will the asset lose power?
   - 0.0: If the asset has no threat of losing commercial power because it uses no electricity or has sufficient backup generation on-site to last until power is restored.
   - 0.5: If the asset could/will lose commercial power and there is some backup generation on-site to last a short amount of time but would not be sufficient if commercial power is not restored quickly.
   - 1.0: If the asset will lose commercial power and there is no backup generation on-site. The asset will remain inoperable until commercial power is restored.
2. Is the asset constructed using flood-resistant materials (concrete, ceramic, pressure-treated lumber)?
   - 0.0: The asset is constructed using flood-resistant materials at least up to the expected flood levels.
   - 0.5: The asset is partially constructed using flood-resistant materials, requiring some repairs after the event.

144

- 1.0: The asset is not constructed using flood-resistant materials, and significant repairs would be needed if flooding reached expected levels.

3. Is the asset sealed so that water cannot enter (dry flood-proofed, backflow valves, etc.)?
   - 0.0: The asset is sealed to prevent water from entering the asset up to the expected flood levels.
   - 0.5: The asset is sealed to prevent water from entering, but not fully up to expected flood levels.
   - 1.0: The asset is not sealed to prevent water from entering.

4. Are electrical system components (including generators/fuel) raised above expected flood levels?
   - 0.0: All electrical components are located above the expected flood level, therefore no damage to electrical components is expected.
   - 0.5: Some electrical components are expected to be damaged from the flood.
   - 1.0: At least most of the electrical components are located at a height that would render them susceptible to damage from the flood.

5. Is HVAC equipment located above expected flood levels?
   - 0.0: No HVAC equipment exists at the asset, or all HVAC equipment is located above the expected flood level, therefore no damage to the HVAC system is expected.
   - 0.5: Some HVAC equipment is expected to be damaged from the flood.
   - 1.0: At least most of the HVAC equipment is at a height that would render it susceptible to damage from the flood.

6. Are all fuel/chemical storage tanks and other components that could be washed away or float in a flood properly anchored or secured?
   - 0.0: All components that could be washed away or float during the flood event have been properly anchored or secured.

- 0.5: Some noncritical components could be lost during the flood event because they are not properly anchored or secured.
- 1.0: Critical components would be lost during the flood event because they are not properly anchored or secured.

7. Are spare parts or critical equipment inventory available for use if repairs are required due to the flood event?
   - 0.0: Spare parts are kept in inventory for all critical components that may be damaged during the flood event.
   - 0.5: Spare parts are kept in inventory for most critical components that may be damaged during the flood event.
   - 1.0: Few spare parts for critical equipment are kept in inventory.

8. Are Standard Operating Procedures (SOPs) or Emergency Response Plans (ERPs) developed to prepare for responding to a flood event?
   - 0.0: SOPs or ERPs are developed for the flood event and are regularly updated and exercised.
   - 0.5: SOPs or ERPs are developed for the flood event but may be outdated and are not exercised.
   - 1.0: SOPs or ERPs are not developed for flood events.

Once these questions are evaluated, the average of the values determines the overall vulnerability of the asset to flooding for that level flood. If any questions are not applicable to the asset type, then take the average of the remaining questions.

SECTION D.4: THREAT LIKELIHOOD

The floods being assessed are due to the 100-year storm and the 500-year storm. This allows the threat likelihoods to be set as the anticipated recurrence interval for each of those storms. The threat likelihoods for each level storm are as follows:

- 100-year storm = 1/100 = 0.01
- 500-year storm = 1/500 = 0.002

SECTION D.5: EXAMPLE

- Asset: Pump station
- Located in floodplain? Yes, for both 100- and 500-year storms
- Expected fatalities and serious injuries: None

100-year Storm:

Consequences

C(100):

Repair and replacement cost: $50,000

Cleanup cost: $10,000

Lost revenue due to service denial: $0

Community economic losses: $0

C(100) = $50,000 + $10,000 + $0

C(100) = $60,000

Vulnerability

V(100):

Question	Ans.
(1) Will the asset lose power?	0.5
(2) Is the asset constructed using flood-resistant materials (concrete, ceramic, pressure-treated lumber)?	0
(3) Is the asset sealed so that water cannot enter (dry flood-proofed, backflow valves, etc.)?	0
(4) Are electrical system components (including generators/fuel) raised above expected flood levels?	1
(5) Is HVAC equipment located above expected flood levels?	0.5
(6) Are all fuel/chemical storage tanks and other components that could be washed away or float in a flood properly anchored or secured?	0
(7) Are spare parts or critical equipment inventory available for use if repairs are required due to the flood event?	1

Question	Ans.
(8) Are Standard Operating Procedures (SOPs) or Emergency Response Plans (ERPs) developed to prepare for responding to a flood event?	0.5
Average (V(100))	0.44

Threat Likelihood

$T(100) = 0.01$

Risk

$R(100) = C(100) \times V(100) \times T(100)$

$R(100) = \$60{,}000 \times 0.44 \times 0.01$

$R(100) = \$264$

500-year Storm:

Consequences

C(500):

Repair and replacement cost: $75,000

Cleanup cost: $20,000

Lost revenue due to service denial: $10,000

Community economic losses: $0

$C(500) = \$75{,}000 + \$20{,}000 + \$10{,}000$

$C(500) = \$105{,}000$

Vulnerability

V(500):

Question	Ans.
(1) Will the asset lose power?	0.5
(2) Is the asset constructed using flood-resistant materials (concrete, ceramic, pressure-treated lumber)?	0
(3) Is the asset sealed so that water cannot enter (dry flood-proofed, backflow valves, etc.)?	0.5
(4) Are electrical system components (including generators/fuel) raised above expected flood levels?	1

Question	Ans.
(5) Is HVAC equipment located above expected flood levels?	1
(6) Are all fuel/chemical storage tanks and other components that could be washed away or float in a flood properly anchored or secured?	0
(7) Are spare parts or critical equipment inventory available for use if repairs are required due to the flood event?	1
(8) Are Standard Operating Procedures (SOPs) or Emergency Response Plans (ERPs) developed to prepare for responding to a flood event?	0.5
Average (V(500))	0.56

Threat Likelihood

$T(500) = 0.002$

Risk

$R(500) = C(500) \times V(500) \times T(500)$

$R(500) = \$105,000 \times 0.56 \times 0.002$

$R(500) = \$118$

Total Risk

$Risk = \$264 + \$118$

$Risk = \$382$

This page intentionally blank.

APPENDIX E

Hurricane

SECTION E.1:   INTRODUCTION

A hurricane is a severe tropical cyclone with sustained winds of 74 mph or greater. Hurricanes and tropical storms have the potential to cause a great deal of damage to drinking water and wastewater utilities due to heavy rainfall and inland flooding, coastal storm surge, and high winds.

Typical impacts that may lead to service interruptions include, but are not limited to:

- pipe breaks due to washouts, uprooted trees, etc., which could result in sewage spills or low water pressure throughout the service area;
- loss of power and communication infrastructure due to high winds;
- combined sewer overflows (CSOs) due to flooding;
- restricted access to facilities and collection and distribution system assets due to debris and floodwaters;
- loss of water quality testing capability during the storm due to restricted facility and laboratory access and damage to utility equipment; and
- system contamination.

The Saffir-Simpson Hurricane Wind Scale (Table E.1) is a 1 to 5 categorization based on the hurricane's intensity. The scale provides examples of the type of damage and impacts in the United States associated with winds of the indicated intensity. The scale does not address the potential for other hurricane-related impacts, such as storm surge, rainfall-induced floods, and tornadoes.

Earlier versions of this scale incorporated central pressure and storm surge as components of the categories. The central pressure was used as a proxy for the winds as accurate wind speed intensity measurements from aircraft reconnaissance were not routinely

151

available for hurricanes until 1995. Storm surge was also quantified by category in the earliest published versions of the scale.

However, hurricane size (extent of hurricane-force winds), local bathymetry (depth of near-shore waters), topography, and the hurricane's forward speed and angle to the coast also affect the surge that is produced. For example, the very large Hurricane Ike in 2008 made landfall in Texas as a Category 2 hurricane and had peak storm surge values of about 20 ft. In contrast, tiny Hurricane Charley struck Florida in 2004 as a Category 4 hurricane and produced a peak storm surge of only about 7 ft. These storm surge values were substantially outside of the ranges suggested in the original scale.

To help reduce public confusion about the impacts associated with the various hurricane categories as well as to provide a more scientifically defensible scale, the storm surge ranges, flooding impact, and central pressure statements were removed from the scale and only peak winds are used in the revised version.

During the open public comment period for the draft of the Saffir-Simpson Hurricane Wind Scale, many people suggested that the National Weather Service (NWS) develop a storm-surge-specific scale as well as improve its forecasting of storm surge. There are some researchers who advocate developing another scale for hurricanes specifically geared toward storm surge impact by incorporating aspects of the system's size.

However, the National Hurricane Center does not believe that such scales would be helpful or effective at conveying the storm surge threat. For example, if 2008's Hurricane Ike had made landfall in Palm Beach, Fla., the resulting storm surge would have been 8 ft, as opposed to the 20 ft that occurred where Ike made landfall on the upper Texas coast. These greatly differing surge impacts arise from differences in the local bathymetry (the shallow Gulf waters off Texas enhance storm surge, while the deep ocean depths off southeastern Florida inhibit surge). The proposed storm surge scales that consider storm size do not consider these local factors that play a crucial role in determining actual surge impacts.

Along the coast, storm surge is often the greatest threat to life and property from a hurricane. In the past, large death tolls have resulted

Table E.1 Saffir-Simpson Hurricane Wind Scale

Category	Winds (1-min sustained winds, mph)	Summary	People, Livestock, and Pets	Apartments, Shopping Centers, and Industrial Buildings	High-Rise Windows and Glass	Signage, Fences, and Canopies	Trees	Power and Water
1	74–95	Very dangerous winds will produce some damage.	People, livestock, and pets struck by flying or falling debris could be injured or killed.	Some apartment building and shopping center roof coverings could be partially removed. Industrial buildings can lose roofing and siding especially from windward corners, rakes, and eaves. Failures to overhead doors and unprotected windows will be common.	Windows in high rise buildings can be broken by flying debris. Falling and broken glass will pose a significant danger even after the storm.	There will be occasional damage to commercial signage, fences, and canopies.	Large branches of trees will snap, and shallow-rooted trees can be toppled.	Extensive damage to power lines and poles will likely result in power outages that could last a few to several days.
2	96–110	Extremely dangerous winds will cause extensive damage.	There is a substantial risk of injury or death to people, livestock, and pets due to flying and falling debris.	There will be a substantial percentage of roof and siding damage to apartment buildings and industrial buildings. Unreinforced masonry walls can collapse.	Windows in high rise buildings can be broken by flying debris. Falling and broken glass will pose a significant danger even after the storm.	Commercial signage, fences, and canopies will be damaged and often destroyed.	Many shallowly rooted trees will be snapped or uprooted and block numerous roads.	Near-total power loss is expected with outages that could last from several days to weeks. Potable water could become scarce as filtration systems begin to fail.
3	111–129	Devastating damage will occur.	There is a high risk of injury or death to people, livestock, and pets due to flying and falling debris.	There will be a high percentage of roof covering and siding damage to apartment buildings and industrial buildings. Isolated structural damage to wood or steel framing can occur. Complete failure of older metal buildings is possible, and older unreinforced masonry buildings can collapse.	Numerous windows will be blown out of high-rise buildings resulting in falling glass, which will pose a threat for days to weeks after the storm.	Most commercial signage, fences, and canopies will be destroyed.	Many trees will be snapped or uprooted, blocking numerous roads.	Electricity and water will be unavailable for several days to a few weeks after the storm passes.

(continued)

153

Table E.1   Saffir-Simpson Hurricane Wind Scale (*Continued*)

Category	Winds (1-min sustained winds, mph)	Summary	People, Livestock, and Pets	Apartments, Shopping Centers, and Industrial Buildings	High-Rise Windows and Glass	Signage, Fences, and Canopies	Trees	Power and Water
4	130–156	Catastrophic damage will occur.	There is a very high risk of injury or death to people, livestock, and pets due to flying and falling debris.	There will be a high percentage of structural damage to the top floors of apartment buildings. Steel frames in older industrial buildings can collapse. There will be a high percentage of collapse of older, unreinforced masonry buildings.	Most windows will be blown out of high-rise buildings resulting in falling glass, which will pose a threat for days to weeks after the storm.	Nearly all commercial signage, fences, and canopies will be destroyed.	Most trees will be snapped or uprooted and power poles downed. Fallen trees and power poles will isolate residential areas.	Power outages will last for weeks to possibly months. Long-term water shortages will increase human suffering. Most of the area will be uninhabitable for weeks or months.
5	157 or higher	Catastrophic damage will occur.	People, livestock, and pets are at very high risk of injury or death from flying or falling debris, even if indoors in mobile homes or framed homes.	Significant damage to wood roof commercial buildings will occur due to loss of roof sheathing. Complete collapse of many older metal buildings can occur. Most unreinforced masonry walls will fail, which can lead to the collapse of the buildings. A high percentage of industrial buildings and low-rise apartment buildings will be destroyed.	Nearly all windows will be blown out of high-rise buildings resulting in falling glass, which will pose a threat for days to weeks after the storm.	Nearly all commercial signage, fences, and canopies will be destroyed.	Nearly all trees will be snapped or uprooted and power poles downed. Fallen trees and power poles will isolate residential areas.	Power outages will last for weeks to possibly months. Long-term water shortages will increase human suffering. Most of the area will be uninhabitable for weeks or months.

Source: National Hurricane Center, 2019

from the rise of the ocean associated with many of the major hurricanes that have made landfall. Hurricane Katrina (2005) is a prime example of the damage and devastation that can be caused by surge. At least 1,500 people lost their lives during Katrina, and many of those deaths occurred, directly or indirectly, because of storm surge.

Storm surge is an abnormal rise of water generated by a storm, over and above the predicted astronomical tides. Storm surge should not be confused with storm tide, which is defined as the water level rise due to the combination of storm surge and the astronomical tide. This rise in water level can cause extreme flooding in coastal areas particularly when storm surge coincides with normal high tide, resulting in storm tides reaching up to 20 ft or more in some cases.

The maximum potential storm surge for a particular location depends on several factors. Storm surge is a complex phenomenon because it is sensitive to the slightest changes in storm intensity, forward speed, size (radius of maximum winds or RMW), angle of approach to the coast, central pressure (minimal contribution compared to the wind), and the shape and characteristics of coastal features such as bays and estuaries.

Other factors that can impact storm surge are the width and slope of the continental shelf. A shallow slope will potentially produce a greater storm surge than a steep shelf. For example, a Category 4 storm hitting the Louisiana coastline, which has a wide and shallow continental shelf, may produce a 20-ft storm surge, while the same hurricane in a place such as Miami Beach, Fla., where the continental shelf drops off quickly, might see an 8- or 9-ft surge.

Adding to the destructive power of surge, battering waves may increase damage to buildings directly along the coast. Water weighs approximately 1,700 lbs/yd^3; extended pounding by frequent waves can demolish any structure not specifically designed to withstand such forces. Additionally, currents created by tides combine with the waves to severely erode beaches and coastal highways. Buildings and buried infrastructure such as pipelines that survive hurricane winds can be damaged if their foundations are undermined and weakened by erosion.

The Sea, Lake, and Overland Surges from Hurricanes (SLOSH) model is a computerized numerical model developed by the NWS to

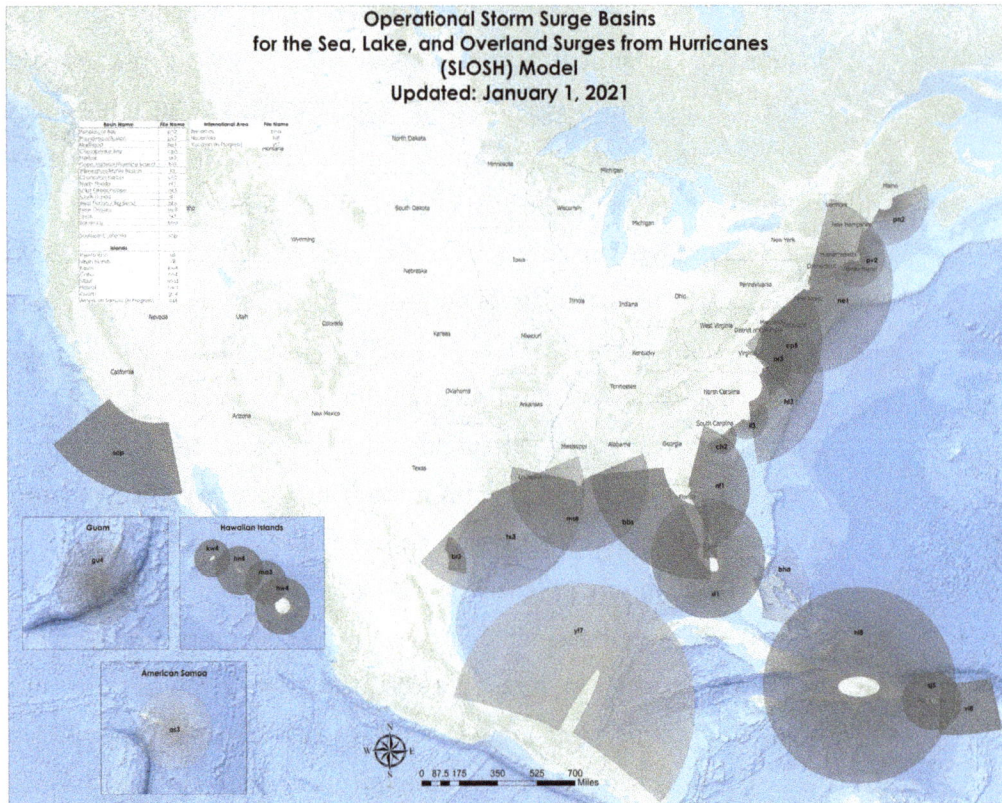

Source: National Weather Service/National Oceanic and Atmospheric Administration

Figure E.1 Operational storm surge basins for the Sea, Lake, and Overland Surges from Hurricanes (SLOSH) model

estimate storm surge heights resulting from historical, hypothetical, or predicted hurricanes by considering the atmospheric pressure, size, forward speed, and track data. These parameters are used to create a model of the wind field, which drives the storm surge.

SLOSH has been applied to the entire US Atlantic and Gulf of Mexico coastlines. Coverage also extends to Hawaii, Puerto Rico, Virgin Islands, and the Bahamas. As shown in Figure E.1, the SLOSH model coverage is subdivided into 32 regions or basins. These basins are centered upon particularly susceptible features: inlets, large coastal centers of population, low-lying topography, and ports. An example of a typical computational domain, or basin, is the New Orleans basin.

SECTION E.2:   CONSEQUENCES

The following sections provide an example for the calculation of risk related to a hurricane event. Use of this calculation is nonmandatory, and other calculation methods may be considered.

Components of consequence to be estimated for hurricane include:

- repair/replacement costs wind damage,
- repair replacement costs due to storm surge damage,
- fatalities/serious injuries,
- owner's revenue loss due to service denial, and
- costs associated with purchasing, shipping, and distributing bottled water.

Sec. E.2.1   Repair/Replacement Costs Due to Wind Damage

In estimating the consequence of wind damage from hurricanes, note that most natural hazards do not result in destruction of the assets they encounter. Rather, partial damage is incurred, so repair and restoration are more frequent than replacement. For example, experience has shown that piping systems are quite robust and will survive a seismic event in most cases. The piping systems used in chemical plants and refineries are generally well-supported, welded systems constructed of ductile metals. It is assumed that large, heavy-walled vessels will be reusable. The cost is primarily the repair and replacement of the plant equipment. To maintain comparability, general damage factors are provided in Table E.2.

The consequence component associated with wind damage is determined by selecting the appropriate damage factor from Table E.2 and multiplying that by the repair and replacement cost of the asset. For example, the replacement cost of a pump station designed to UBC code may be $500,000. The consequence associated with hurricane wind damage would be $500,000 \times 0.5 = \$250,000$.

Sec. E.2.2   Repair/Replacement Costs Due to Storm Surge

The damage associated with storm surge would result from water inundation and wave action. The National Hurricane Center Storm Surge Risk Maps provide inundation heights for coastal areas for each category hurricane; the maps can be found at https://experience.arcgis.com/experience/203f772571cb48b1b8b50fdcc3272e2c. Flood loss is

Table E.2 Damage factors of selected equipment types

Damage Factors	Asset Types and Mountings
0.2	• Slab-mounted equipment—pumps, valves, compressors, meters, electric motors, electrical controls, consoles, etc.
	• Ground-level storage tanks
	• Water heaters and similar equipment equipped with seismic restraints
	• Automobiles and trucks, heavy equipment
0.3	• Aboveground piping designed to accepted codes and standards such as ANSI B31.1, and ANSI B31.3
	• Pressure vessels designed to ASME codes and standards
0.4	• Elevated storage tanks
0.5	• Buildings designed to UBC Code or equivalent
0.75	• Buildings not designed to codes
1.00	• Portable buildings and trailers

assumed to consist primarily of severe electrical damage to wiring and motors, switchgear, and telephone and communication equipment; residual mud and debris; mold; rot; and damage to carpets, drapes, furniture, and equipment that is sensitive to oxidation (rusting). For applications where replacement is not warranted, repair costs associated with contaminated electrical wiring and switchgear may include cleaning impacted assets with purified water, drying appropriately, and then rewiring that includes applying anticorrosive lubricant.

Locate the asset on the storm surge map and compare the asset's grade and equipment elevation to the estimated inundation height for the applicable hurricane category. Estimate the repair/replacement costs for damaged equipment not above the flood level.

This process is repeated for each category hurricane.

Sec. E.2.3 Fatalities and Serious Injuries

Fatalities and serious injuries associated with hurricane impacts to water system assets should be estimated for each category hurricane. The fatalities and serious injuries caused by high winds should be assigned to wind risk; similarly, fatalities and serious injuries caused by storm surge should be assigned to storm surge risk. Do not double count fatalities and serious injuries.

Sec. E.2.4 Owner's Revenue Loss

The owner's revenue loss is associated with operational outages resulting in service denial and lost revenue. For each category hurricane, estimate the down time and resulting service denial.

Operational outages should be assigned to either wind risk or storm surge risk, but not both, to avoid double counting. Outages caused by high winds should be applied to wind risk; similarly, outages caused by storm surge should be applied to storm surge risk. If there is anticipated damage attributed to both wind and storm surge, assign the outage to the risk that would result in the longest outage. For example, if wind damage would result in an outage of 3 days, and storm surge would result in an outage of 5 days, assign 5 days of outage to the storm surge risk.

Sec. E.2.5 Cost for Bottled Drinking Water

The owner may be required or encouraged to purchase, ship, and distribute bottled water. Depending on the national event, an agency such as FEMA or the National Guard may be given the mission to provide bottled water.

SECTION E.3: VULNERABILITY

The vulnerability of an asset to hurricanes will vary based on hurricane category level and whether consequences are associated with high winds and/or storm surge. Vulnerability for hurricane winds and storm surge is discussed as follows.

Sec. E.3.1 Wind Vulnerability

To determine the vulnerability of an asset due to high winds, first determine the design wind speed of the asset. If this cannot be determined, use the Uniform Building Code (UBC) minimum wind speed map (see example in Figure E.2) to estimate the most likely design wind speed for structures in the region. It is assumed that structures and equipment designed in accordance with the UBC, which includes most, if not all, critical infrastructure, do not suffer damage unless there is a hurricane that exceeds the design wind speed for that region or a tornado.

159

Using the Saffir-Simpson Hurricane Wind Scale in Table E.1, determine the hurricane category that aligns with the asset's design wind speed. For example, a design wind speed of 110 mph aligns with a Category 2 hurricane.

Assets subjected to hurricanes at or below the design wind speed for the structure are assumed to have a wind vulnerability of 0. If the hurricane is one category above the design wind speed, the wind vulnerability is assumed to be 0.5. If it is two or more categories above the design wind speed, the vulnerability is assumed to be 1.0.

Sec. E.3.2 Storm Surge Vulnerability

Vulnerability to storm surge should be estimated through expert elicitation on a scale of 0 to 1, where 0 assumes zero likelihood that the estimated consequences would occur and 1 assumes a 100% likelihood that estimated consequences would occur. The following questions should be considered to determine the storm surge vulnerability:

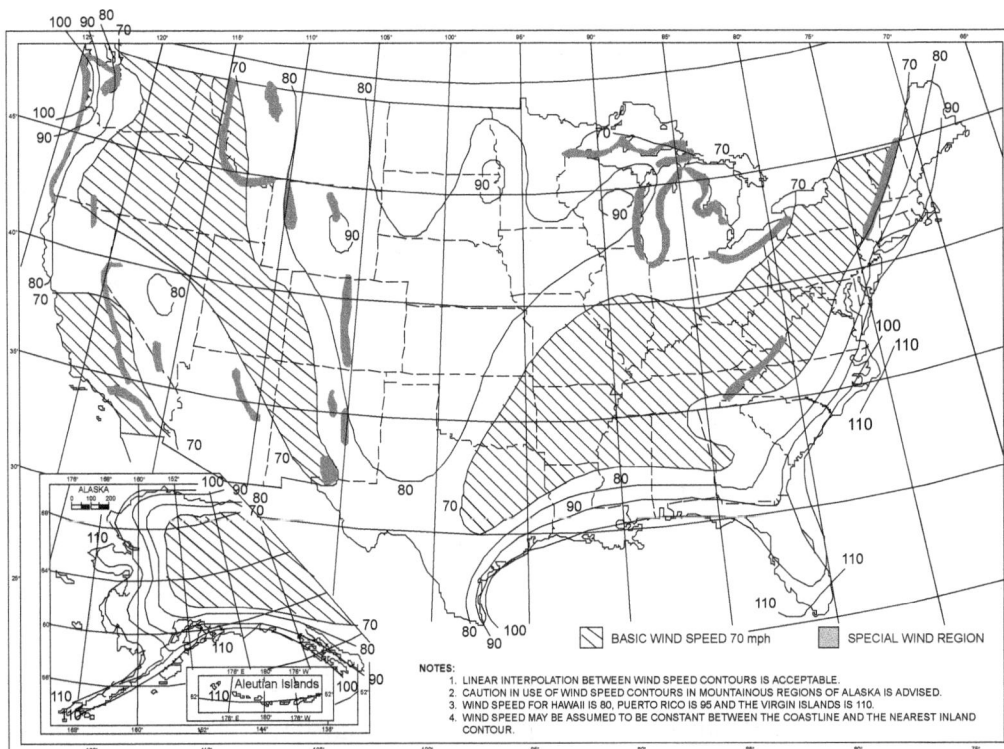

Source: Uniform Building Code, 1997

Figure E.2 Basic wind speed, 50-year recurrence interval

160

- Is there a Standard Operating Procedure (SOP) or Emergency Response Plan (ERP) for preparing for and responding to a flood event? Has it been exercised?
- Will staff be able to access the asset during storm surge conditions? If not, can the asset be operated remotely?
- Are there permanent flood protection measures (e.g., flood wall, flood gates) in place that protect at the adequate level?
- Are there temporary flood protection measures (e.g., flood bags, bastion products, inflated protective barriers) in place that protect at the adequate level?
- Is the asset constructed using flood-resistant materials (e.g., concrete, ceramic, pressure-treated lumber)?
- Is the asset sealed so that water cannot enter (e.g., dry flood-proofed, backflow valves)?
- Are electrical system components (including generators/fuel) located above expected storm surge levels?
- Are critical components located above expected storm surge levels?
- Are there measures in place to minimize damage from wave action?
- Will the asset lose power?
- Are all fuel/chemical storage tanks and other components properly anchored or secured?
- Are spare parts or critical equipment inventory available for use if repairs are required due to the event?

SECTION E.4: THREAT LIKELIHOOD

The threat likelihood of a hurricane is the same whether one is evaluating wind risk or storm surge risk; however, the likelihood changes for each hurricane category. Hurricane frequency data is available from USEPA's Storm Surge Inundation Map Tool: https://epa.maps.arcgis.com/apps/MapSeries/index.html?appid=852ca645500d419e8c6761b923380663.

On the Hurricane Frequency tab, select a map grid to see a table of past hurricane strikes, categorized by the Saffir-Simpson Hurricane Wind Scale (Category 1 through Category 5). (Note: If a table does not pop up when clicking a grid, try using a different internet browser.) The number of strikes can be used to calculate the threat likelihood. For example, 9 strikes between 1900 and 2020 would equate to an annual frequency of 9 strikes/120 years = 0.075.

SECTION E.5: EXAMPLE

Sec. E.5.1 Risk Calculation Overview

The risk associated with hurricanes can be attributed to high winds and/or storm surge. While wind damage may be minimal from low-category hurricanes, storm surge could be significant depending on an asset's location and elevation and the lunar tide cycle. If an asset is not located in an area subject to storm surge, the storm surge risk is assumed to be 0.

Because an asset's vulnerability to high winds can vary greatly from its vulnerability to storm surge, the two components must be calculated separately. Therefore, the hurricane risk is calculated by adding the risk associated with high winds to the risk associated with storm surge, and total hurricane risk is calculated by adding the risk for all five categories.

$$R_{\text{Hurr Cat1}} = R_{\text{Wind Cat1}} + R_{\text{Surge Cat1}} \text{ [repeated for each hurricane category]}$$

$$R_{\text{Total Hurr}} = R_{\text{Hurr Cat1}} + R_{\text{Hurr Cat2}} + R_{\text{Hurr Cat3}} + R_{\text{Hurr Cat4}} + R_{\text{Hurr Cat5}}$$

Table E.3 provides a worksheet for performing this analysis. Further discussion of the components used to calculate risk are provided in Sections E.2–E.4.

Sec. E.5.2 Example of Hurricane Risk Analysis

This section provides an example of a hurricane risk calculation for a water asset. Assume the following for this example:

- Location: Miami, Fla.
- Assume the asset in question is a pump station built to UBC Code with a replacement cost of $2.5M.

See Table E.4 for a completed table calculating hurricane risk for this example. Step-by-step instructions are provided after the table.

Table E.3 Estimation of hurricane risk

Item		Hurricane Category				
		Cat 1	Cat 2	Cat 3	Cat 4	Cat 5
Wind Risk						
Consequences due to High Winds						
(a)	Expected wind damage to asset ($)[1]					
(b)	Fatalities due to wind ($) = number of fatalities × statistical value of life[2]					
(c)	Serious injuries due to wind ($) = number of serious injuries × statistical value of serious injuries[2]					
(d)	Lost revenue ($) due to wind					
(e)	Total consequence due to high winds = (a + b + c + d)					
Wind Vulnerability						
(f)	Wind vulnerability (0, 0.5, 1) (dependent on asset wind speed design basis)[3]					
Threat Likelihood						
(g)	Threat likelihood for hurricane category (occurrences/year)[4]					
Wind Risk						
(h)	Wind risk = (e) × (f) × (g)					
Storm Surge Risk						
Consequences due to Storm Surge						
(i)	Expected storm surge damage to asset ($)[5]					
(j)	Fatalities due to storm surge ($) = number of fatalities × statistical value of life[2]					
(k)	Serious injuries due to storm surge ($) = number of serious injuries × statistical value of serious injuries[2]					
(l)	Lost revenue ($) due to storm surge					
(m)	Total consequence due to storm surge = (i + j + k + l)					
Storm Surge Vulnerability						
(n)	Storm surge vulnerability[3]					
Threat Likelihood						
(o)	Threat likelihood for hurricane category (occurrences/year)[4]					
Storm Surge Risk						
(p)	Storm surge risk ($) = (m) × (n) × (o)					
Total Hurricane Risk						
(q)	Hurricane risk ($) = (h) + (p)					
(r)	**Total hurricane risk =** annual risk for Cat 1 + annual risk for Cat 2 + annual risk for Cat 3 + annual risk for Cat 4 + annual risk for Cat 5					

NOTES:
[1] Refer to Table E.2 in Sec. E.2.1 for wind damage factors.
[2] Value of a statistical life can be obtained from US Department of Transportation guidance.
[3] Refer to Sec. E.3 for determination of vulnerability.
[4] Refer to Sec. E.4 and https://epa.maps.arcgis.com/apps/MapSeries/index.html?appid=852ca645500d419e8c6761b923380663 for determination of threat likelihood.
[5] Refer to Sec. E.2.2 and https://experience.arcgis.com/experience/203f772571cb48b1b8b50fdcc3272e2c for damage due to storm surge.

Table E.4 Example calculation of hurricane risk

Item		Hurricane Category				
		Cat 1	Cat 2	Cat 3	Cat 4	Cat 5
Wind Risk						
Consequences due to High Winds						
(a)	Expected wind damage to asset ($)[1]	$1,250,000	$1,250,000	$1,250,000	$1,250,000	$1,250,000
(b)	Fatalities due to wind ($) = number of fatalities × statistical value of life[2]	$0	$0	$0	$0	$0
(c)	Serious injuries due to wind ($) = number of serious injuries × statistical value of serious injuries[2]	$0	$0	$0	$0	$0
(d)	Lost revenue ($) due to wind	$0	$50,000	$150,000	$250,000	$500,000
(e)	Total consequence due to high winds = (a + b + c + d)	$1,250,000	$1,300,000	$1,400,000	$1,500,000	$1,750,000
Wind Vulnerability						
(f)	Wind vulnerability (0, 0.5, 1) (dependent on asset wind speed design basis)[3]	0	0	0.5	1.0	1.0
Threat Likelihood						
(g)	Threat likelihood for hurricane category (occurrences/year)[4]	0.066	0.066	0.066	0.074	0.017
Wind Risk						
(h)	Wind risk = (e) × (f) × (g)	$0	$0	$46,200	$111,000	$29,750
Storm Surge Risk						
Consequences due to Storm Surge						
(i)	Expected storm surge damage to asset ($)[5]	$0	$0	$10,000	$100,000	$100,000
(j)	Fatalities due to storm surge ($) = number of fatalities × statistical value of life[2]	$0	$0	$0	$0	$0
(k)	Serious injuries due to storm surge ($) = number of serious injuries × statistical value of serious injuries[2]	$0	$0	$0	$0	$0
(l)	Lost revenue ($) due to storm surge	$0	$0	See Note 6	See Note 6	See Note 6
(m)	Total consequence due to storm surge = (i + j + k + l)	$0	$0	$10,000	$100,000	$100,000
Storm Surge Vulnerability						
(n)	Storm surge vulnerability[3]	0.9	0.9	0.9	0.9	0.9
Threat Likelihood						
(o)	Threat likelihood for hurricane category (occurrences/year)[4]	0.066	0.066	0.066	0.074	0.017
Storm Surge Risk						
(p)	Storm surge risk ($) = (m) × (n) × (o)	$0	$0	$594	$6,660	$1,530
Total Hurricane Risk						
(q)	Hurricane risk ($) = (h) + (p)	$0	$0	$46,794	$117,660	$31,280
(r)	**Total Hurricane Risk =** annual risk for Cat 1 + Cat 2 + Cat 3 + Cat 4 + Cat 5	**$195,734 / year**				

NOTES:
[1] Refer to Table E.2 in Sec. E.2.1 for wind damage factors.
[2] Value of a statistical life can be obtained from US Department of Transportation guidance.
[3] Refer to Section E.3 for determination of vulnerability.
[4] Refer to Section E.4 and https://epa.maps.arcgis.com/apps/MapSeries/index.html?appid=852ca645500d419e8c6761b923380663 for determination of threat likelihood.
[5] Refer to Sec. E.2.2 and https://experience.arcgis.com/experience/203f772571cb48b1b8b50fdcc3272e2c for damage due to storm surge.
[6] For this example, the lost revenue due to wind damage is greater than damage due to storm surge. To avoid double counting, the lost revenue was assigned to wind risk only.

Calculating Hurricane Risk Associated with Wind

Wind Consequence

1. From Table E.2, find that the damage factor for a building designed to UBC Code is 0.5. Therefore, the wind consequence is $2.5M × 0.5 = $1,250,000. This is applied to all five hurricane categories. (*Row a of Table E.4*)

2. Assume no fatalities or serious injuries associated with the water system for any of the five hurricane categories. (*Rows b and c of Table E.4*)

3. Estimate the outage severity (MGD) and duration (days) due to wind damage for the various level categories and calculate corresponding lost revenue.
   - Cat 1: $0
   - Cat 2: $50,000
   - Cat 3: $150,000
   - Cat 4: $250,000
   - Cat 5: $500,000

If there is anticipated damage attributed to both wind and storm surge, assign the outage/lost revenue to the risk that would result in the longest outage. (*Row d of Table E.4*)

4. Add together the components of wind consequence. (*Row e of Table E.4*)

Wind Vulnerability

1. Referencing Figure E.2, the wind design basis for Miami, Fla., is 110 mph.

2. From Table E.1 (Saffir-Simpson Hurricane Wind Scale), a Category 2 hurricane would be expected to have wind speeds up to 110 mph. Therefore, Category 3 and greater hurricanes are of concern since they would exceed the design basis loading. As a result, the wind vulnerability is as follows (*Row f of Table E.4*):
   - Cat 1 and 2: 0
   - Cat 3: 0.5
   - Cat 4, and 5: 1.0

165

Table E.5 Hurricane strike frequency table from USEPA tool for Miami, Fla., for 1900–2021

Hurricane Statistics	
Category	**Count of Strikes**
All	35
1	8
2	8
3	8
4	9
5	2

Wind Threat Likelihood

Referencing the USEPA Storm Surge Inundation Map Tool (https://epa.maps.arcgis.com/apps/MapSeries/index.html?appid=852ca 645500d419e8c6761b923380663), determine the return period (threat likelihood) for each category hurricane for Miami, Fla. (*Row g of Table 4*)

Using Table E.5, calculate the threat likelihood for each hurricane category in Miami:

- Category 1: 8 / 121 = 0.066/year
- Category 2: 8 / 121 = 0.066/year
- Category 3: 8 / 121 = 0.066/year
- Category 4: 9 / 121 = 0.074/year
- Category 5: 2 / 121 = 0.017/year

Wind Risk

Multiply Wind Consequence, Vulnerability, and Threat Likelihood. (*Row h of Table E.4*)

- Category 1: $1,250,000 × 0 × 0.066 = $0
- Category 2: $1,300,000 × 0 × 0.066 = $0
- Category 3: $1,400,000 × 0.5 × 0.066 = $46,200
- Category 4: $1,500,000 × 1.0 × 0.074 = $111,000
- Category 5: $1,750,000 × 1.0 × 0.017 = $29,750

Calculating Hurricane Risk Associated with Storm Surge

Storm Surge Consequence

1. Locating the asset on the National Hurricane Center Storm Surge Risk Map, the asset is subject to storm surge at the following levels:

- No storm surge for Category 1 hurricane
- Less than 3 ft for Category 2 hurricane
- Between 3 and 6 ft for Category 3 hurricane
- Between 6 and 9 ft for Category 4 hurricane
- Over 9 ft for Category 5 hurricane

Because the asset has electrical equipment located at 6 ft, damage is expected from storm surge associated with Category 4 and 5 hurricanes (assume $100,000). Storm surge associated with a Category 3 hurricane would result in minimal damage to the asset (assume $10,000). (*Row i of Table E.4*)

2. Assume no fatalities or serious injuries associated with the water system for any of the five hurricane categories. (*Rows j and k of Table E.4*)

3. Estimate the outage severity (MGD) and duration (days) due to storm surge for the various level categories and calculate corresponding lost revenue. Based on estimated damage for this example, assume $150,000 in lost revenue for Category 4 and 5 hurricanes, no lost revenue for Category 3 and below.
   - Category 1, 2, 3: $0
   - Category 4 and 5: $150,000

If there is anticipated lost revenue attributed to both wind and storm surge, assign the outage/lost revenue to the risk that would result in the longest outage. In this example, lost revenue due to wind is higher; therefore, lost revenue is not included in storm surge consequence. (*Row l of Table E.4*)

4. Add together the components of storm surge consequence. (*Row m of Table E.4*)

Storm Surge Vulnerability

Estimate the asset's vulnerability to storm surge using the questions in Sec. E.3.2. In this example, the vulnerability is assumed to be the same for all five hurricane categories (0.9), but it can be different based on expected storm surge height and existing flood mitigation measures.

Storm Surge Threat Likelihood

Storm surge threat likelihood is the same as wind threat likelihood. (*Row o of Table E.4*)

167

Storm Surge Risk

Multiply Storm Surge Consequence, Vulnerability, and Threat Likelihood. (*Row p of Table E.4*)

- ○ Category 1: $0 × 0.9 × 0.066 = $0
- ○ Category 2: $0 × 0.9 × 0.066 = $0
- ○ Category 3: $10,000 × 0.9 × 0.066 = $594
- ○ Category 4: $100,000 × 0.9 × 0.074 = $6,660
- ○ Category 5: $100,000 × 0.9 × 0.017 = $1,530

Total Hurricane Risk

1. Add together Wind Risk and Storm Surge Risk to calculate Hurricane Risk for each hurricane category. (*Row q of Table E.4*)

- ○ Category 1: $0 + $0 = $0
- ○ Category 2: $0 + $0 = $0
- ○ Category 3: $46,200 + $594 = $46,794
- ○ Category 4: $111,000 + $6,600 = $117,660
- ○ Category 5: $29,750 + $1,530 = $31,280

2. Add together Hurricane Risk for all five categories. (*Row r of Table E.4*)

- • **Total Hurricane Risk** = $0 + $0 + $46,794 + $117,660 + $31,280 = **$195,734**

SECTION E.6   REFERENCES

International Conference of Building Officials. 1997. *Uniform Building Code*.

National Hurricane Center. 2019. *Saffir-Simpson Hurricane Wind Scale*.

National Oceanic and Atmospheric Administration. (n.d.). *Sea, Lake, and Overland Surges from Hurricanes (SLOSH)*. www.nhc.noaa.gov/surge/slosh.php

APPENDIX F

Ice Storms

SECTION F.1: INTRODUCTION

Although there may be a variety of consequences from an ice storm (e.g., damage to aboveground equipment, impassable roads), this approach assumes that the major impact on the water sector is the loss of electrical power. The vulnerability assesses the likelihood that the threat will succeed to inflict the maximum reasonable damage. For the case of loss of power, the countermeasure to reduce vulnerability and consequences is the capacity of on-site generation to meet the minimum electricity requirement to keep the utility operating at an acceptable level until commercial power is restored.

The Sperry-Piltz Ice Accumulation Index (SPIA®Index) (Table F.1) is a forward-looking, ice accumulation and ice damage prediction index that uses an algorithm of researched parameters that, when combined with National Weather Service forecast data, predicts the projected footprint, total ice accumulation, and resulting potential damage from approaching ice storms. Note: The SPIA®Index proprietary algorithm does NOT include "observed ice" accumulations in its formulary at this time; however, such observed ice measurements may be ingested into the proprietary algorithm at a later time.

It is a tool to be used for risk management and/or winter weather preparedness. The SPIA®Index is to ice storms what the Enhanced Fujita Scale is to tornadoes and what the Saffir-Simpson Scale is to hurricanes. Previous to this hazard scale development, no such forward-looking ice accumulation and ice damage index had been utilized to predict—days in advance—the potential damage to overhead utility systems, along with outage duration possibilities, from freezing rain and/or ice storm events.

The SPIA®Index uses three key parameters to formulate the algorithm:

- storm total rainfall, converted to ice accumulation;
- wind; and

Table F.1   The Sperry-Piltz Ice Accumulation Index or "SPIA®Index" (©February 2009)

Ice Damage Index	Average NWS Ice Amount (in in.) *Revised-October 2011	Sustained Winds (mph)	Damage and Impact Descriptions (*Modified January 2019)
0	< 0.25	<15	Minimal risk of damage to exposed utility systems; no alerts or advisories needed for crews, few outages.
1	010–0.25	15–25	Some isolated or localized utility outages possible, typically lasting only a few hours. Roads/Bridges may become slick and hazardous. Some tree limb damage.
1	0.25–0.50	<15	Some isolated or localized utility outages possible, typically lasting only a few hours. Roads/Bridges may become slick and hazardous. Some tree limb damage.
2	0.10–0.25	25–35	Numerous utility interruptions expected, typically lasting 24 to 72 hours. Roads/ Travel conditions may be extremely hazardous. Moderate tree damage expected.
2	0.25–0.50	15–25	Numerous utility interruptions expected, typically lasting 24 to 72 hours. Roads/ Travel conditions may be extremely hazardous. Moderate tree damage expected.
2	0.50–0.75	<15	Numerous utility interruptions expected, typically lasting 24 to 72 hours. Roads/ Travel conditions may be extremely hazardous. Moderate tree damage expected.
3	0.10–0.25	>35	Widespread utility outages with damage to main feeder lines and equipment expected. Tree limb damage is excessive. Outages may last from 3 to 5 days.
3	0.25–0.50	25–35	Widespread utility outages with damage to main feeder lines and equipment expected. Tree limb damage is excessive. Outages may last from 3 to 5 days.
3	0.50–0.75	15–25	Widespread utility outages with damage to main feeder lines and equipment expected. Tree limb damage is excessive. Outages may last from 3 to 5 days.
3	0.75–1.00	<15	Widespread utility outages with damage to main feeder lines and equipment expected. Tree limb damage is excessive. Outages may last from 3 to 5 days.
4	0.25–0.50	>35	Prolonged and widespread utility interruptions with extensive damage to main distribution feeder lines and some high voltage transmission lines/structures. Outages lasting 5 to 10 days.
4	0.50–0.75	25–35	Prolonged and widespread utility interruptions with extensive damage to main distribution feeder lines and some high voltage transmission lines/structures. Outages lasting 5 to 10 days.
4	0.75–1.00	15–25	Prolonged and widespread utility interruptions with extensive damage to main distribution feeder lines and some high voltage transmission lines/structures. Outages lasting 5 to 10 days.
4	1.00–1.50	<15	Prolonged and widespread utility interruptions with extensive damage to main distribution feeder lines and some high voltage transmission lines/structures. Outages lasting 5 to 10 days.
5	0.50–0.75	>35	Catastrophic damage to entire exposed utility systems, including both distribution and transmission networks. Outages could last several weeks in some areas. Shelters needed.
5	0.75–1.00	>25	Catastrophic damage to entire exposed utility systems, including both distribution and transmission networks. Outages could last several weeks in some areas. Shelters needed.
5	1.00–1.50	>15	Catastrophic damage to entire exposed utility systems, including both distribution and transmission networks. Outages could last several weeks in some areas. Shelters needed.
5	> 1.50	Any	Catastrophic damage to entire exposed utility systems, including both distribution and transmission networks. Outages could last several weeks in some areas. Shelters needed.

(Categories of damage are based upon combinations of precipitation totals, temperatures, and wind speeds/directions.)

Source: Courtesy of Sidney K. Sperry, President and CEO, SPIDI Technologies, LLC, Guthrie, Okla.

- temperatures during the event period.

These parameters, when used in conjunction with digital forecasts from local NWS Weather Forecast Offices, have been shown to accurately predict the duration, intensity, and damage capability of ice storms.

SECTION F.2: CONSEQUENCES

Consequences from an ice storm can include serious injuries, fatalities, service outage, financial loss to the owners, and/or economic loss to the community. However, the major consequence from an ice storm for the water sector is the resultant loss of power, which this method will focus on. If power is lost for a significant amount of time, then the utility will be unable to meet its demand with resulting loss of income. The backup or alternate power capacity of a utility determines its vulnerability to the threat of an ice storm. The days of lost income sustained by the utility as a result of a Category 1 or greater ice storm can be expressed as:

$$C(X) = (CR \times DF \times (t_D + t_R)) + (LL(X) \times SVL) + (SI(X) \times SVSI)$$

If the statistical value of life and serious injury is not used, then the equation for consequence is:

$$C(X) = CR \times DF \times (t_D + t_R)$$

Where:

X = level of ice storm from SPIA$^©$ Index

C(X) = the total financial consequences for an X-level ice storm

CR = utility commodity rate in \$/MG

DF = daily flow through the asset in MG

t_D = the maximum number of days of lost commercial power for an X-level storm

t_R = the number of recovery days needed to achieve pre-ice storm operating levels after utility recovers its power supply

LL(X) = number of fatalities for an X-level ice storm

SI(X) = number of serious injuries for an X-level ice storm

SVL = Statistical Value of Life

SVSI = Statistical Value of Serious Injury

The values for t_D are listed in Table F.2.

These values coincide with the maximum power outage for all applicable levels in the SPIA$^©$ Index. Since this approach assumes that ice storms will produce relatively little physical damage to water sector infrastructure, the utility's consequence of a given ice storm is the lost revenue for each day that the plant is out of operation, the number or value of lives lost, and the number or value of serious injuries.

Table F.2 t_D Values by storm level

SPIA© Index Level	t_D Value
1	3 h (0.125 days)
2	1 day
3	5 days
4	10 days
5	21 days

SECTION F.3:   VULNERABILITY

The estimated consequence of ice storms is the loss of power from the power utility. The vulnerability is found using two variables: the probability of an outage falling between the days specified in the SPIA© Index and the number of days of alternative power (on-site generation) available to the water utility.

The values found for the probability that a utility will be unable to provide power for each SPIA© Index level $P_p(X)$ are:

The equation for vulnerability of a facility to an ice storm is:

$$V(X) = \left(\frac{t_D - t_B}{t_D} \right) \times P_p(X)$$

Where:

$V(X)$ = the vulnerability of a facility to an X-level ice storm

t_D = the maximum number of days of lost commercial power for an X-level ice storm

t_B = the number of days of backup power available on-site at the utility

$P_p(X)$ = the probability that the electric provider will be down for the specified number of days for an X-level ice storm

Table F.3 P_p Values by storm level

SPIA© Index Level	P_p Value	Outage Range Covered
1	0.32	< 3 h
2	0.149	3 h–1 day
3	0.497	1–5 days
4	0.0349	5–10 days
5	0.000000354 (3.54×10^{-7})	≥ 10 days

If V(X) is negative, then a value of zero should be used for the risk calculation. This means there is no vulnerability to the storm. If V(X) is greater than one, it should be treated as one because probability of occurrence cannot be greater than 1.0 or 100%.

SECTION F.4:   THREAT LIKELIHOOD

The threat likelihood of an ice storm occurring within one year is a result of two factors: wind speed and ice thickness. As identified in the SPIA$^©$ Index, the combination of these two occurrences determines the intensity of the storm. The probabilities that specified ice thicknesses and wind speed will occur have been determined for each range specified in the SPIA$^©$ Index. The SPIA$^©$ Index combines the ice thickness and wind speed to 1) determine the appropriate damage index for a given event, and 2) calculate the probability that an ice thickness and a wind speed will occur at the same place, at the same time.

$$P \ (I \ AND \ W) = P(I) \times P(W)$$

Where:

I = ice thickness (in.) at a given location

W = wind speed (mph) at a given location

P(I) = the probability that an ice thickness within a given range will occur at a given location

P(W) = the probability that a wind speed within a given range will occur at a given location

The possible ice thickness ranges and wind speeds as used by the SPIA$^©$ are:

To obtain the probability of any level storm, the sum of the possible combinations for that level is calculated. The equations to calculate the likelihoods for each level of ice storms are as follows. (Note: Level 5 probability must accommodate all cases that exceed Level 4.)

$$T(1) = [P(0.1 \leq I < 0.25) \times P(15 < W < 25)] + [P(0.25 \leq I < 0.5) \times P(W < 15)]$$

$$T(2) = [P(0.1 \leq I < 0.25) \times P(25 < W < 35)] + [P(0.25 \leq I < 0.5) \times P(15 < W < 25)] + [P(0.5 \leq I < 0.75) \times P(W < 15)]$$

Table F.4a Possible ice thickness ranges

Ice Thickness (in.)
$0.1 \leq I < 0.25$
$0.25 \leq I < 0.5$
$0.5 \leq I < 0.75$
$0.75 \leq I < 1.0$
$1.0 \leq I < 1.5$
$I \geq 1.5$

Table F.4b Possible wind speed ranges

Wind Speed (mph)
$W < 15$
$15 \leq W < 25$
$25 \leq W < 35$
$W \geq 35$

$$T(3) = [P(0.1 \leq I < 0.25) \times P(W \geq 35)] + [P(0.25 \leq I < 0.5) \times P(25 < W < 35)] + [P(0.5 \leq I < 0.75) \times P(15 < W < 25)] + [P(0.75 \leq I < 1.0) \times P(W < 15)]$$

$$T(4) = [P(0.25 \leq I < 0.5) \times P(W \geq 35)] + [P(0.5 \leq I < 0.75) \times P(25 < W < 35)] + [P(0.75 \leq I < 1.00) \times P(15 < W < 25)] + [P(1.00 \leq I < 1.5) \times P(W < 15)]$$

$$T(5) = [P(0.5 \leq I < 0.75) \times P(W \geq 35)] + \{P(0.75 \leq I < 1.00) \times [P(25 < W < 35) + P(W \geq 35)]\} + \{P(1.00 \leq I < 1.5) \times [P(15 < W < 25) + P(25 < W < 35) + P(W \geq 35)]\} + \{P(I \geq 1.5) \times [P(W < 15) + P(15 < W < 25) + P(25 < W < 35) + P(W \geq 35)]\}$$

OR

$$T(X) = \sum (P(I \text{ AND } W))$$

Where:

$T(X)$ = the threat likelihood of an X-level ice storm

$P(I \text{ AND } W)$ = the probability of an ice thickness and a wind speed occurring on the same day

If the utility is in a part of the country that does not experience ice storms, then the threat likelihood will be zero, resulting in a risk for ice storms of zero.

Table F.5 Probability that ice accumulation of thickness—I (in inches)—will occur in the geographical area (P(I))

States	P(0.1 ≤ I <0.25)	P(0.25 ≤ I <0.5)	P(0.5 ≤ I < 0.75)	P(0.75 ≤ I <1.0)	P(1.0 ≤ I <1.5)	P(I ≥ 1.5)
CO, DE, MD, NC, ND, NE, SC, SD, VA, WV	0.139	0.277	0.225	0.113	0.042	0.001
KS, NM, OK, TX	0.092	0.181	0.183	0.148	0.147	0.031
IA, MI, MN, WI	0.101	0.164	0.136	0.094	0.081	0.016
IL, IN, KY, MO, OH	0.059	0.405	0.438	0.086	0.003	0
CT, MA, ME, NH, NJ, NY, PA, RI, VT	0.095	0.168	0.156	0.121	0.121	0.03
AL, AR, FL, GA, LA, MS, TN	0.074	0.216	0.277	0.22	0.14	0.007
AK, AZ, CA, HI, ID, MT, NV, OR, UT, WA, WY	0.143	0.226	0.155	0.074	0.030	0.001

Use Table F.5 to identify the probabilities that each range of ice thicknesses will occur in the asset location.*

Next, the probability of a wind speed occurring is found. The probabilities of occurrence for each wind range of the SPIA$^©$ Index are given in the following maps (Figures F.1–F.4). A value for each map may be determined in one of two ways: by using the average value for the area the asset is in, or by interpolating the value based on the minimum and maximum values for the shaded area the asset is in.

SECTION F.5:   FINAL RISK CALCULATION

In a J100 analysis, risk is defined as the product of the consequences, vulnerability, and threat probability (R = C × V × T). Using the approach suggested here, the risk to a utility for ice storms in the United States can be determined. This equation is repeated for each SPIA$^©$ Index-level ice storm, and then summed to determine the total risk.

---

* Changnon, S.A. 2003. "Characteristics of Ice Storms in the United States." *J. Appl. Meteor.*, 42(5)630–639. http://journals.ametsoc.org/doi/abs/10.1175/1520-0450(2003)042%3C0630%3ACOISIT%3 E2.0.CO%3B2

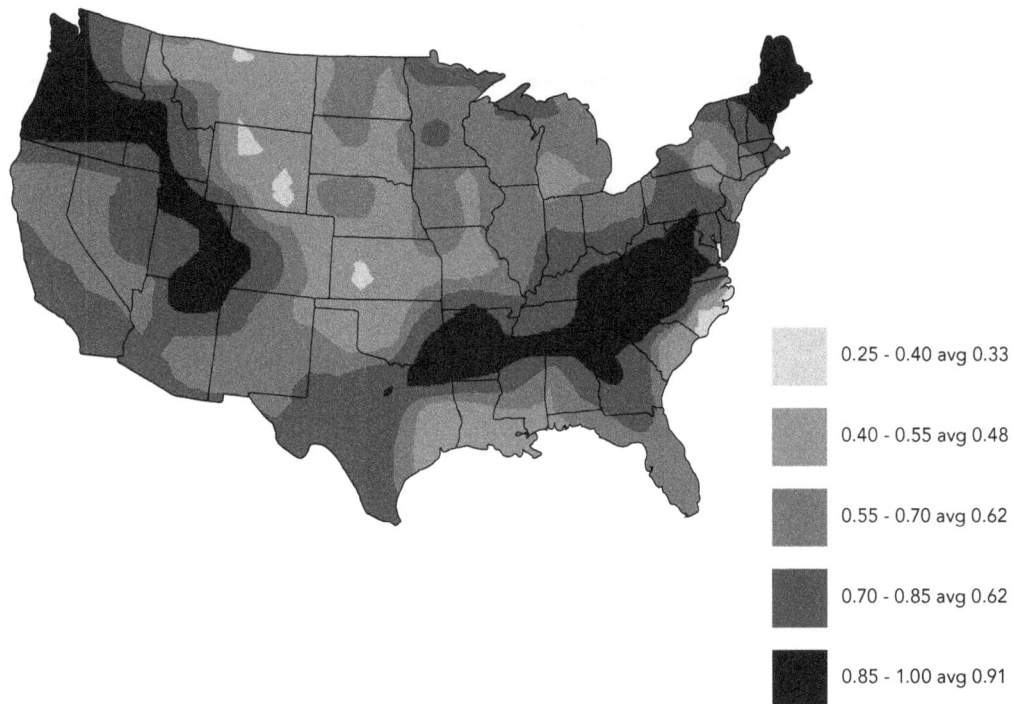

	0.25 - 0.40 avg 0.33
	0.40 - 0.55 avg 0.48
	0.55 - 0.70 avg 0.62
	0.70 - 0.85 avg 0.62
	0.85 - 1.00 avg 0.91

Figure F.1   Wind speed probabilities for W <15 mph

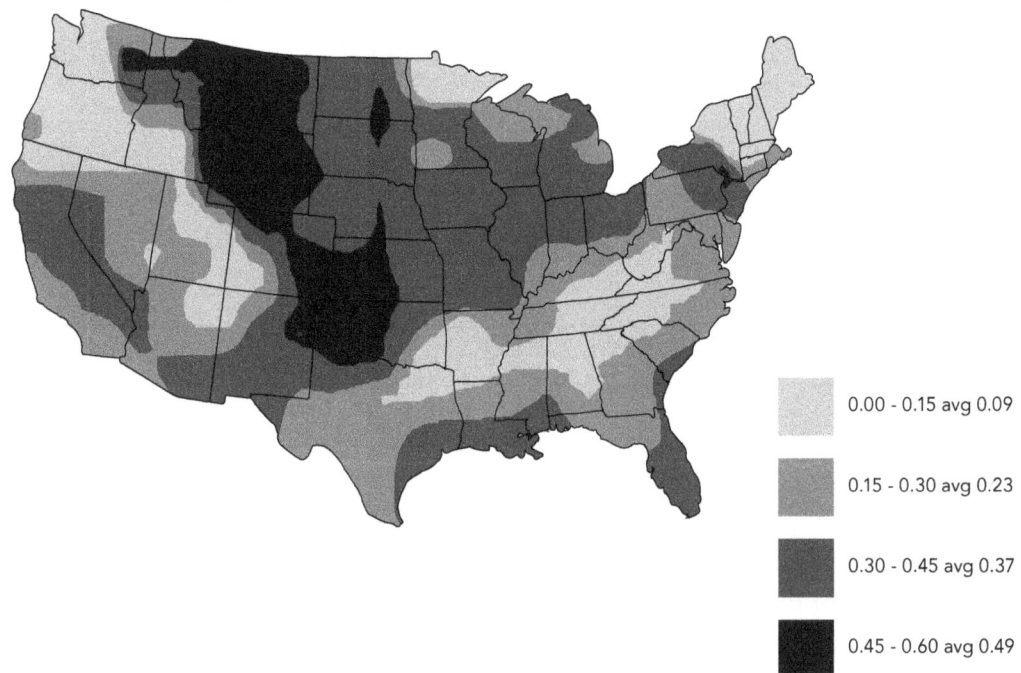

	0.00 - 0.15 avg 0.09
	0.15 - 0.30 avg 0.23
	0.30 - 0.45 avg 0.37
	0.45 - 0.60 avg 0.49

Figure F.2   Wind speed probabilities for 15 mph < W <25 mph

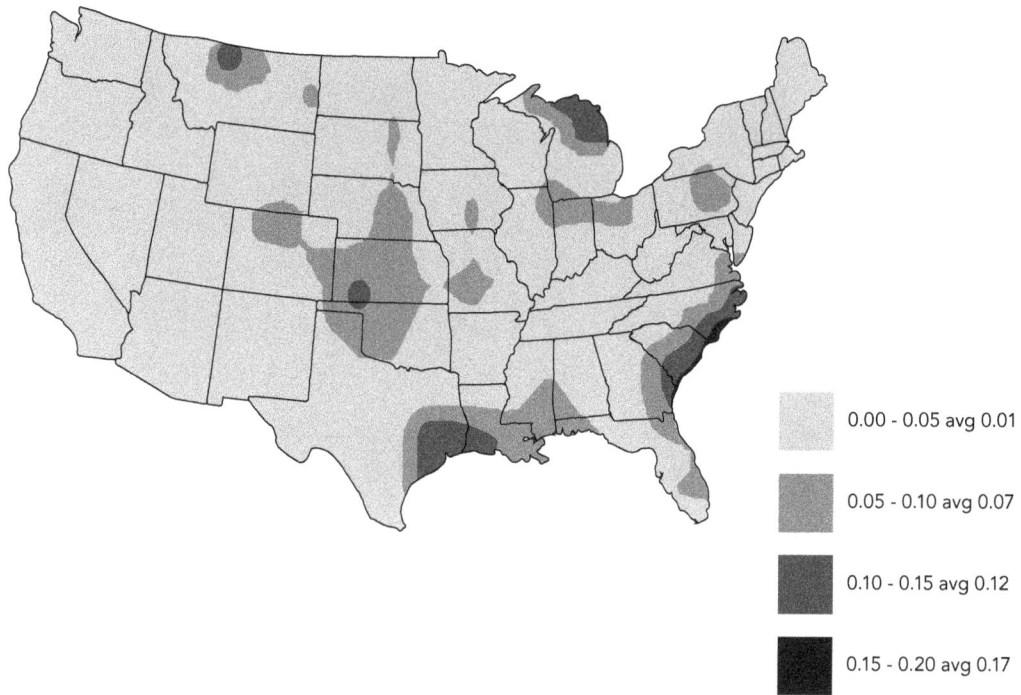

Figure F.3   Wind speed probabilities for 25 mph < W <35 mph

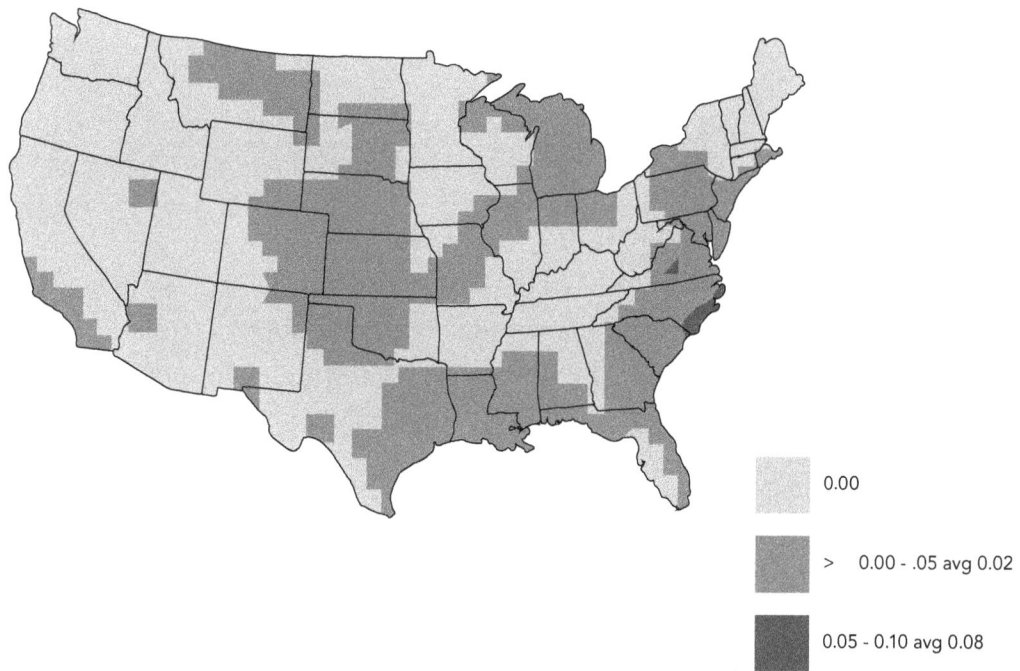

Figure F.4   Wind speed probabilities for W >35 mph

$$R = \sum_{X=1}^{5} (C(X) \times V(X) \times T(X))$$

Where:

$$C(X) = \left(CR \times DF \times (t_D + t_R)\right) + \left(LL(X) \times SVL\right) + \left(SI(X) \times SVSI\right)$$

$$V(X) = \left(\frac{t_D - t_B}{t_D}\right) x\ P_p(X)$$

$$T(X) = \Sigma\left(P(I\ AND\ W)\right)$$

SECTION F.6: EXAMPLE

Assumptions:

- City: Chantilly, Va., Fairfax County
- Latitude: 38.9°
- Longitude: −77.45°
- Days of backup power available on-site (t_B) = 4 days
- Recovery Time (t_R):
  - Level 1: 0 days
  - Level 2: 0 days
  - Level 3: 0.5 days
  - Level 4: 3 days
  - Level 5: 6 days
- Commodity Rate (CR) = \$2,500/MG
- Daily Flow: 2 MG
- Expected fatalities and serious injuries: 0

The probability of ice thickness formation in a specific range in Virginia (from Table F.5):

P(I):

- $0.1 \leq I < 0.25 = 0.139$
- $0.25 \leq I < 0.5 = 0.277$
- $0.5 \leq I < 0.75 = 0.225$
- $0.75 \leq I < 1.0 = 0.113$
- $1.0 \leq I < 1.5 = 0.042$
- $I \geq 1.5 = 0.001$

The probability of average wind speeds in Virginia (from Figures 1–4):

P(W):

- $W < 15 = 0.698$
- $15 \leq W < 25 = 0.258$
- $25 \leq W < 35 = 0.035$
- $W \geq 35 = 0.008$

Calculated Risk of a Level 1 Ice Storm:

Consequences

$C(1) = (CR \times DF \times (t_D + t_R))$

$C(1) = \$2,500 \times 2MG \times (0.125 \text{ days} + 0 \text{ days})$

C(1) = \$625

Vulnerability

$V(1) = [(t_D - t_B) / t_D] \times P_p$

$V(1) = [(0.125 - 4) / 0.125] \times 0.32$

V(1) = 0 (Default for results less than 0)

Threat Likelihood

$T(1) = [P(0.1 \leq I < 0.25) \times P(15 < W < 25)] + [P(0.25 \leq I < 0.5) \times P(W < 15)]$

$T(1) = (0.139 \times 0.258) + (0.277 \times 0.698)$

T(1) = 0.229

Risk

$R(1) = C(1) \times V(1) \times T(1)$

$R(1) = 625 \times 0 \times 0.229$

R(1) = \$0

Calculated Risk of a Level 2 Ice Storm:

Consequences

$C(2) = (CR \times DF \times (t_D + t_R))$

$C(2) = \$2,500 \times 2 MG \times (1 \text{ day} + 0 \text{ days})$

C(2) = \$5,000

Vulnerability

$V(2) = [(t_D - t_B) / t_D] \times P_p$

$V(2) = [(1 - 4) / 1] \times 0.149$

V(2) = 0 (Default for results less than 0)

Threat Likelihood

$T(2) = [P(0.1 \leq I < 0.25) \times P(25 < W < 35)] + [P(0.25 \leq I < 0.5) \times P(15 < W < 25)] + [P(0.5 \leq I < 0.75) \times P(W < 15)]$

$T(2) = (0.139 \times 0.035) + (0.277 \times 0.258) + (0.225 \times 0.698)$

$T(2) = 0.234$

Risk

$R(2) = C(2) \times V(2) \times T(2)$

$R(2) = \$5{,}000 \times 0 \times 0.234$

$R(2) = \$0$

Calculated Risk of a Level 3 Ice Storm:

Consequences

$C(1) = (CR \times DF \times (t_D + t_R))$

$C(1) = \$2{,}500 \times 2 \ MG \times (5 \ days + 0.5 \ days)$

$C(1) = \$27{,}500$

Vulnerability

$V(3) = [(t_D - t_B) / t_D] \times P_p$

$V(3) = [(5 - 4) / 5] \times 0.497$

$V(3) = 0.0994$

Threat Likelihood

$T(3) = [P(0.1 \leq I < 0.25) \times P(W \geq 35)] + [P(0.25 \leq I < 0.5) \times P(25 < W < 35)] + [P(0.5 \leq I < 0.75) \times P(15 < W < 25)] + [P(0.75 \leq I < 1.0) \times P(W < 15)]$

$T(3) = (0.139 \times 0.008) + (0.277 \times 0.035) + (0.225 \times 0.258) + (0.113 \times 0.698)$

$T(3) = 0.148$

Risk

$R(3) = C(3) \times V(3) \times T(3)$

$R(3) = \$27{,}500 \times 0.0994 \times 0.148$

$R(3) = \$405$

Calculated Risk of a Level 4 Ice Storm:

Consequences

$C(1) = (CR \times DF \times (t_D + t_R))$

$C(1) = \$2,500 \times 2 \text{ MG} \times (10 \text{ days} + 3 \text{ days})$

$C(1) = \$65,000$

Vulnerability

$V(4) = [(t_D - t_B) / t_D] \times P_p$

$V(4) = [(10 - 4) / 10] \times 0.0349$

$V(4) = 0.0209$

Threat Likelihood

$T(4) = [P(0.25 \leq I <0.5) \times P(W \geq 35)] + [P(0.5 \leq I <0.75) \times P(25 < W < 35)] + [P(0.75 \leq I <1.00) \times P(15 < W < 25)] + [P(1.00 \leq I < 1.5) \times P(W < 15)]$

$T(4) = (0.277 \times 0.008) + (0.225 \times 0.035) + (0.113 \times 0.258) + (0.042 \times 0.698)$

$T(4) = P(4) = 0.069$

Risk

$R(4) = C(4) \times V(4) \times T(4)$

$R(4) = \$65,000 \times 0.0209 \times 0.069$

$R(4) = \$94$

Calculated Risk of a Level 5 Ice Storm:

Consequences

$C(1) = (CR \times DF \times (t_D + t_R))$

$C(1) = \$2,500 \times 2MG \times (21 \text{ days} + 6 \text{ days})$

$C(1) = \$135,000$

Vulnerability

$V(5) = [(t_D - t_B) / t_D] \times P_p$

$V(5) = [(21 - 4) / 21] \times 3.54 \times 10^{-7}$

$V(5) = 2.866 \times 10^{-7}$

Threat Likelihood

$T(5) = [P(0.5 \leq I < 0.75) \times P(W \geq 35)] + \{P(0.75 \leq I < 1.00) \times [P(25 < W < 35) + P(W \geq 35)]\} + \{P(1.00 \leq I < 1.5) \times [P(15 < W < 25) + P(25 < W < 35) + P(W \geq 35)]\} + \{P(I \geq 1.5) \times [P(W < 15) + P(15 < W < 25) + P(25 < W < 35) + P(W \geq 35)]\}$

$T(5) = (0.225 \times 0.008) + [0.113 \times (0.035 + 0.008)] + [0.042 \times (0.258 + 0.035 + 0.008)] + [0.001 \times (0.698 + 0.258 + 0.035 + 0.008)]$

$T(5) = 0.020$

Risk

$R(5) = C(5) \times V(5) \times T(5)$

$R(5) = \$135,000 \times 2.866 \times 10^{-7} \times 0.020$

$R(5) = \$0.00$

Calculated Total Risk of an Ice Storm:

Total Risk

$$R = \sum_{X=1}^{5} (C(X) \times V(X) \times T(X))$$

$R = \$0 + \$0 + \$405 + \$94 + \$0$

$R = \$499$

APPENDIX G

Earthquake Risk

SECTION G.1:   INTRODUCTION

Sec. G.1.1 Overview

Earthquake risk varies greatly based on location, so for many utilities, a basic screening should be done to gauge the level of risk posed by earthquakes. Systems at greater risk would benefit from a relatively in-depth and rigorous analysis, while those with minimal risk might apply a more cursory analysis. Even systems with relatively low risk might benefit from attention to earthquakes in the design of new facilities, since at the point of initial construction, some mitigations are so inexpensive that they can be justified even when risk is low. However, assessing risk is not the same as designing an important facility, and the information presented here should not be used for design. A design process incorporates specific steps, usually in compliance with a specific building code, and is done by an engineer with specific training and licensing. By contrast, a high-level review of an existing system for seismic risk is much less detailed and seeks not to definitively analyze any specific facility but rather to understand how seismic concerns may (or may not) impact a utility's overall risk.

A basic screening for risk can be done using a seismic risk map. USGS no longer provides such maps, but the USGS site has links to services run by others—some of which are available at low or no cost. For example, the Structural Engineers Association of California (SEAOC)[*] and Applied Technology Council (ATC)[†] both provide free interactive online maps that show seismic hazards and other hazards such as wind and snow.

To use one of the available online maps as a screening tool, it's necessary to understand a few of the industry terms used in earthquake risk assessment as well as the ways an earthquake may damage a

---

[*] https://seismicmaps.org

[†] https://hazards.atcouncil.org

water system. Basic concepts and terminology are given subsequently, followed by a discussion of earthquake consequences in Sec. G.2. Sec. G.3 discusses asset vulnerability. Sec. G.4 discusses earthquake likelihood and describes risk screening based on likelihood. Sec. G.5 provides an example risk assessment calculation.

Sec. G.1.2 Terminology of Earthquake Analysis

To effectively use seismic screening tools, a knowledge of some basic terminology as found in building codes and similar publications is needed. The terminology changes frequently, creating a potentially confusing lexicon of ever-shifting jargon. New terms are frequently introduced for old concepts, sometimes with subtle but important differences. A few of the most important terms are discussed as follows.

Risk category.   This is a new term appearing in codes circa 2012, replacing the term *occupancy category*, and effectively supplanting stand-alone use of *importance factors*. A *risk category* is a number from 1 to 4, usually expressed in Roman I to IV, to denote how important a facility is, with Risk Category I indicating an unimportant and Risk Category IV a very important facility. Risk category may be said to "replace" an earlier concept called the *importance factor* because by assigning a *risk category*, one is implicitly assigning an "importance factor"—a factor that multiplies assumed (seismic) loads on a structure by a factor ranging from 1.0 for Category I to 1.5 for Category IV. A full discussion is beyond the scope of this guide, but it should be noted that water tanks are called out as Risk Category IV, while pipelines may reasonably be placed into Categories II, III, or IV, depending on their importance.

While a conservative risk category is aimed at improving the chances of a facility's survival in an earthquake, simply assigning Risk Category IV during design does not guarantee that a facility will be operable.

Probabilistic Seismic Hazard Analysis (PSHA).   PSHA is a specific, fairly complex, analytical method that is appropriately applied to specific facilities rather than to geographically dispersed systems. The method involves assessing the risk to a facility of all known seismic sources, considering both frequent smaller events as well as rare larger events, and providing essentially a single set of design criteria

184

that encompasses the aggregate risk. PSHA is *not* recommended for assessments of system-wide vulnerability since it would not reflect the reality that any single earthquake would not imperil all facilities in a large system to the same degree. PSHA would thus overstate risk and fail to reveal the benefits of practical mitigations such as redundancy across a distributed system, e.g., multiple sources of supply and redundant piping. Instead of PSHA, a system-wide assessment (if needed) should be based on scenario analysis, which will be discussed later.

Recurrence interval. This is the average period of time between earthquakes at or above a given magnitude. Two frequently cited recurrence intervals in building codes are 475 years and 2,475 years. These seemingly odd numbers flow from numbers that are more "round": 475 years is associated with a 10% chance of exceedance in 50 years, and 2,475 years is associated with a 2% chance of exceedance in 50 years.[‡]

For most natural hazards, the code is based on the 475-year recurrence, e.g., a code-compliant building is designed for wind speeds that have a nominal 10% chance of being exceeded during a 50-year design life. Seismic and tsunami loads historically were based on a 475-year recurrence, but starting in 2008, the criteria were revised in most codes to require consideration of a 2,475-year recurrence. The intent was to ensure that areas vulnerable to rare but major earthquakes would address seismic considerations in design of structures. It was recognized that a 2,475-year basis might be very conservative in locales of higher seismicity, so the code has significant additional complexity to reduce the design basis in those areas back to approximately what it was with a 475-year assumption. A key point to understand when comparing seismic codes to other natural hazard codes is the level of performance expected. A code-compliant building survives its wind design event with possibly little or no damage, while that same building need only avoid collapse during its design earthquake. Owners seeking

---

[‡] As it turns out, the numbers 475 and 2,475 commonly shown with up to four significant digits connote false precision. The mathematical formulation, while appealing due to its simplicity, relies upon an assumption that the timing of an event is independent of prior events, i.e., that it's a Poisson process. While reasonably true for some natural hazards, it's not true for earthquakes, as highlighted in "seismic gap" theory.

a higher level of performance by the building may find benefits to additional conservatism beyond code minimums.

Scenario earthquake. A scenario earthquake is a specific earthquake, historic or postulated, that is used to assess risk. A utility should consider a range of credible scenario earthquakes to cover known sources of seismicity and a range of recurrence intervals as discussed previously. The number of scenarios considered, and the degree of precision in the analysis, can be tailored to risk. Table 1 of J100-21 classifies earthquakes in terms of their peak ground acceleration aka PGA (discussed in G.2.1) and suggests that a utility consider "the lowest magnitude [event] … that poses potentially unacceptable risks …" as well as potentially greater events that may be less frequent than that lowest-magnitude event. Following this approach, two bookend scenarios could be considered: that lowest-magnitude scenario as well as the worst-case. For some utilities, a quick screening (as discussed in Threat Likelihood, Sec. G.4.1) may reveal that even the worst-case seismic concerns are minor compared to other risks and can either be ignored entirely for the J100 analysis or analyzed using a single scenario.

SECTION G.2:   CONSEQUENCE

Sec. G.2.1 Earthquake Metrics

Earthquakes damage facilities in two fundamental ways: by shaking them and/or by deforming them. For powerful earthquakes, shaking effects may be felt for up to hundreds of miles from an earthquake's location, while deformation effects are usually limited to a much smaller zone nearer to the earthquake's location (known as its epicenter). While in general shaking is stronger close to the epicenter than at further distances, the drop-off of shaking intensity with distance can vary substantially based on the local geology.

In contrast to *intensity* which varies by location, the *magnitude* of an earthquake is an attribute of the earthquake itself and does not vary by location. Magnitude is usually expressed by those in the seismology field as "moment magnitude" M_w, which is a measure of total energy released in the earthquake, expressed in a logarithmic scale (so an increase of one unit denotes ten times the energy). The very largest earthquakes may exceed M_w 9, but even an M_w 5 can inflict damage.

Unfortunately, there are competing measures of magnitude that are similar enough to cause confusion. As an example, media reports often characterize an earthquake's "magnitude" with no units or sometimes they refer to Richter magnitude. Richter magnitude (also called M_L) is positively correlated with M_w, but the relationship is nonlinear since the two systems are based on different principles. (Richter measures maximum amplitude of seismic wave, adjusted for distance, while M_w measures total energy.)

The intensity of an earthquake felt at a specific location will be more important for a utility owner than the earthquake's absolute magnitude. That intensity, or shaking effect, is usually characterized by the acceleration or velocity of the ground at a location of interest. Two often-used measures are peak ground acceleration (PGA) and peak ground velocity (PGV). In the United States, PGA is usually expressed as a fraction of gravitational acceleration, e.g., 0.5g corresponds to 16.1 ft/s^2 or 4.9 m/s^2. PGV cannot be expressed as a ratio of gravity, so it's expressed in absolute units, and the chosen units vary, e.g., in./s or cm/s.

Ground displacement or deformation comes in two basic forms: transient and permanent. While both types of displacement may matter, the focus is usually on permanent ground deformation (PGD), expressed in linear units, e.g., inches.

The abbreviation PGD may also be used to denote peak ground displacement during an earthquake, which is a different thing than the PGD discussed in this document. Peak ground displacement refers to the maximum (elastic) displacement of the ground during an earthquake. That is different than the PGD in this discussion, which does not arise from elastic oscillation but rather from permanent changes in the ground, such as those caused by landslides, liquefaction, or fault offset. The alternative PGD is mentioned because it can create confusion. And since it's derivable from a seismograph record, people may believe that PGD metrics are widely available. They are not, when one is referring to the permanent ground deformation flavor of PGD.

When assessing a large water system, one may have to proceed with little or no quantification of permanent ground deformation.

Many other measures of intensity exist besides PGA, PGV, and PGD. Some are derivative, such as the ratio of PGA/PGV, PGV^2/PGA, etc. Some are crafted to specific circumstances; for example, "spectral acceleration" is the single value of acceleration that is determined through calculation to produce a response equivalent to a ground motion history for a structure with a specific period of vibration and damping characteristics. Some such as the Modified Mercalli Intensity (MMI) are focused on perceived effects on structures and people, rather than on exposure. And some such as Japan's shindo use different units of measure. This guide will focus on how to obtain and use the three basic metrics PGA, PGV, and PGD.

The metrics PGA, PGV, and PGD can be estimated for a specific scenario earthquake. J100-21 provides guidance on scenarios to be considered, as discussed in Sec. G.1. Generally, a scenario will be most useful if an estimate of its recurrence interval is available, since that recurrence interval can be used to compute T, the annual threat likelihood, in the J100-21 risk formulation $R = C \times V \times T$ (see Sec. G.4 for details on threat likelihood).

Online tools as described in Sec. G.1.1 provide estimates of PGA and are helpful for basic screening. For example, if an online tool indicates that the 2,475-year recurrence PGA level is less than 0.20g, seismic risk is a relatively minor concern and is unlikely to be an important part of a J100-21 analysis.

However, when a quick screening shows that the earthquake risk may *not* be minor, then a utility would look for a full suite of metrics including PGA, PGV, and PGD for one or more scenario earthquakes. Unfortunately, the online tools don't provide estimates of either PGV or PGD, which are both particularly helpful in assessing damage to buried assets. For this reason, consider using scenario earthquakes from the USGS Shakemap library (http://earthquake.usgs.gov/data/shakemap/), from other available sources, or developed for a specific project. Typically, Shakemaps will include PGA and PGV.

Use estimates of PGD if possible, but often estimates will not be available. PGD can occur in the form of:

- landslides—caused by destabilizing susceptible formations;
- liquefaction and associated lateral spread and settlement—caused by shaking of poorly consolidated alluvial soils and fills. Once a soil layer liquefies, it may move laterally downgradient;
- lurching—lateral movement of soil blocks due to strong ground motion;
- settlement (nonliquefaction related)—soil consolidation caused by earthquake vibration; and
- surface fault rupture—caused by the causative fault breaking the surface.

Most of these sources of PGD will not be well quantified or may require major geotechnical or geologic studies to quantify values for even a single location. Therefore, the risk analysis for a typical system should focus more on *whether* there is a risk of PGD at a location rather than on quantifying it, and mitigations should be chosen when possible that are insensitive to the precise quantification of PGD. A further discussion of the dominant sources of PGD, and ways to assess whether it is likely, follows.

In many locations, liquefaction can be very problematic and a significant factor in determining a system's seismic vulnerability; for example, it was a dominant cause of infrastructure failures in the 2010–2011 Christchurch, New Zealand, earthquakes. Liquefaction can occur in alluvial deposits along rivers, lakes, and the ocean, and in areas of fill. Many state geology departments have mapped liquefaction susceptibility, and those data are available in GIS format to help qualitatively characterize the risk. As discussed, this will often be sufficient, and it's suggested that any more detailed studies be selectively targeted at critical facilities that are found with the basic screening tools to have a medium or higher risk of liquefaction.

PGD may also occur when shallow faulting reaches the surface (surface fault rupture). Pipelines crossing these faults will likely fail unless they are specially designed to accommodate the displacement. Many reasonable options do not require quantitative analysis. For existing small-diameter pipelines, the options include acceptance of the risk or measures to permit quick repair such as isolation valves. Should the pipeline be replaced, possibly for reasons unrelated to seismic concerns, that opportunity can be taken to reroute to avoid the hazard or to replace the pipe with a strain-tolerant pipe. Detailed expert analysis of a specific fault crossing could reasonably be reserved for relatively important new or existing pipelines for which the options mentioned previously are insufficient.

Even given the importance of PGD to buried assets, and challenges in its estimation, there is still value in estimating the possibility of earthquake damage absent PGD data, i.e., using either PGA and PGV or PGA alone if no other data are available.

The following sections describe how to use PGA, PGV, and PGD to estimate damage to water system components. That discussion is followed by a discussion of how to assess consequences to the water system resulting from damage to its components, and consequences to society—often the most important part of overall earthquake consequence.

As noted in the discussion, the various methods of estimating damage apply to the water system itself. An earthquake may disrupt other infrastructure sectors such as power and transportation, making a full assessment of societal impact very complex and beyond the scope of this document.

Sec. G.2.2 Estimating Damage to Discrete Facilities

Discrete water system facilities include tanks, pump stations, and other point features, many of which are above ground. Earthquakes affect above ground systems differently than they do buried, and the methods to estimate damage differ accordingly. For aboveground systems, it's common to use "fragility curves" that describe the likelihood of various levels of damage to a facility given a level of exposure as measured by PGA. In the context of J100-21's risk equation $R = C \times V \times T$, the fragility analysis provides an estimate V (vulnerability) to different damage levels. For simplicity, it is recommended that a single damage level ("Moderate") be considered; FEMA's HAZUS software associates this level of damage with downtime of three or more days, which is a reasonable threshold given that a three-day outage would be a typical minimum planning assumption for private citizens or institutions. See Sec. G.3 Vulnerability for details on how vulnerability is estimated.

While fragility curves may be available for metrics other than PGA, for simplicity, it's recommended that PGA be used for most J100-21 analyses, although for buried structures, PGV could be used if available. When multiple metrics are used for a single facility, the highest estimated vulnerability V should be used.

Sec. G.2.3 Estimating Damage to Pipelines

For buried pipeline systems, damage is more strongly associated with PGV and PGD rather than with PGA. Also, in contrast to discrete assets such as tanks and pump stations, for which vulnerability (V) is computed as a probability between 0 and 1.0, linear pipelines are modeled differently. For pipelines, a number of failures per 1,000 ft is estimated, and vulnerability is taken as 1.0.

The number of pipeline failures can be roughly estimated by using PGV from the scenario Shakemaps and (when available) liquefaction/lateral spread PGD derived for the specific area. When PGD data are not available (the typical case), it may be assumed to be 6 in. in areas of liquefaction and zero otherwise.

Approximate relationships between PGV or PGD and expected pipeline failures as a function of pipe material can be found in an ALA document (American Lifelines Alliance, 2001).

For ground shaking, the relationship is:

Repair rate/1,000 ft = $K_1 \times 0.00187 \times PGV$

Where:

K_1 = a constant related to expected performance of different pipe materials (See Table G.1, based on ALA)

PGV = in units of in./s

For PGD, the relationship is:

Repair rate/1,000 ft = $K_2 \times 1.06 \times PGD^{0.319} \times L_S$

Where:

K_2 = a constant related to expected performance of different pipe materials (See Table G.1)

PGD = in inches

L_S = the percentage of the formation that will liquefy to the extent that lateral spreading will occur

191

Table G.1 K_1 and K_2 values for pipeline loss estimates

Material	Joint Type	Soils	Diameter	K_1	K_2
Cast iron	Cement	All	Small	1.00	1.00
Cast iron	Cement	Corrosive	Small	1.40	
Cast iron	Cement	Noncorrosive	Small	0.70	
Cast iron	Rubber gasket	All	Small	0.80	0.80
Cast iron	Mechanical restrained				0.70
Welded steel	Lap-arc welded	All	Small	0.60	
Welded steel	Lap-arc welded	Corrosive	Small	0.90	
Welded steel	Lap-arc welded	Noncorrosive	Small	0.30	
Welded steel	Lap-arc welded	All	Large	0.15	0.15
Welded steel	Rubber gasket	All	Small	0.70	0.70
Welded steel	Screwed	All	Small	1.30	
Welded steel	Riveted	All	Small	1.30	
Asbestos cement	Rubber gasket	All	Small	0.50	0.8
Asbestos cement	Cement	All	Small	1.00	1.00
Concrete w/stl cyl.	Lap-Arc welded	All	Large	0.70	0.60
Concrete w/stl cyl.	Cement	All	Large	1.00	1.00
Concrete w/stl cyl.	Rubber gasket	All	Large	0.80	0.70
PVC – C900, C905	Rubber gasket	All	Small	0.50	0.80
PVC – C909 (1)	Restrained	All	Small	0.15	
Ductile iron	Rubber gasket	All	Small	0.50	0.50
Ductile iron (1)	Restrained joint	All	Small	0.25	
Ductile iron (1)	Seismic joint	All	Small	0.15	
HDPE (1) – C906	Fused	All	Small	0.15	

Source: Table created from two American Lifelines Alliance (ALA) tables
NOTE 1. Based on engineering judgment, not from ALA

Of the failures due to PGV, 80 percent are estimated to be leaks and 20 percent breaks (loss in hydraulic continuity). Of the failures due to PGD, 20 percent are estimated to be leaks and 80 percent breaks.

A GIS can be used to overlay utility piping systems and their associated pipe materials on the PGV and PGD hazard maps. The total number of pipe failures can then be calculated, and areas with high concentrations of pipe failures identified.

Sec. G.2.4   Estimates of Overall Consequence

Having estimated damage to system components such as facilities and pipelines from a scenario earthquake, system functionality can be assessed using several different approaches. First, a workshop setting with system operators and engineers familiar with system operation can be used to assess the impact on the system of losing the assets in question. If outage times for the various assets have been determined, system recovery times can also be developed. Note that this requires an assessment of available repair resources and how repair work would be prioritized.

A second way of assessing system-wide impacts is with quantitative modeling. Depending on the system, an existing system hydraulic model such as EPANET can be used to evaluate impacts of various components being out of service. System breaks could be introduced in the model connectivity and leaks and breaks introduced in the hydraulic demand. For this approach to be meaningful, it's important to have clarity on what question the model is supposed to answer and whether the associated assumptions are realistic. If the model is to represent the system state immediately after a quake, before any isolation valves are operated, then breaks should be modeled and due consideration given to loss of pressure. Conversely, if the model is to represent the system state after isolation valves are operated, the assumption of valve operability should be validated with valve exercising.

For simple (wholesale) systems, spreadsheet-based connectivity models have been built that incorporate the probabilities of each asset being operable. While these models can become very complicated, they do offer an output of a numerical probability of serving an end user.

None of the preceding discussions of system impacts accounts for inter-sector dependencies, e.g., reliance of a pump station upon electric power. Inter-sector dependencies, while beyond the scope of this document, may greatly impact several aspects of earthquake consequence, including serviceability, repair time, and the societal costs associated with service interruption. It is advisable for a utility to contact their power provider to gain an understanding of the likely risks of power outage due to earthquakes.

The societal consequence of system outages (i.e., consequence from loss of service) can be estimated separately from the direct water system consequence. Often, those societal costs will greatly exceed direct costs to the owner. An assessment of the overall system serviceability should be made considering the extent of damage to the source, transmission, treatment, distribution pipelines, distribution storage, and pump stations. The downtime can be estimated considering the extent of damage and the resources available to make repairs. Outage areas can be determined considering whether they have a treated supply and distribution facilities to transport water. Given the area of a potential outage—and thus the number of affected customers—plus an estimated duration of the outage, societal consequence can be assessed.

One metric for doing so is the simple product, i.e., the number of person-outage days. While one could then estimate economic value, any such estimate should be viewed as very approximate. FEMA has for years maintained a dollar estimate of the value of a person-day without water, set at $116/day-person in 2021. In reality, the value of water will vary tremendously—not simply among a population but for individuals within a population over time and in accordance with utility theory. Realistic values may range from zero (e.g., for a homeowner whose house is destroyed and unhabitable) to several times the quoted value (e.g., when needed to maintain life itself or to protect an important structure). Values clearly vary over time as well: A single day without water is annoying while a second day may be life-threatening. As a general rule, the marginal utility of water is so variable as to render constant-value assessments unreliable for guiding investment decisions either before or after an earthquake. Even FEMA has published information documenting a factor of ten difference between high and low values. The simplistic constant-dollar assessments are nonetheless shown in this guide because they do illustrate the basic concept, well established by research, that societal costs of a water service disruption tend to be much larger than direct costs to the water utility. Also, the constant-dollar method is still used by FEMA to evaluate grant applications.

Other metrics, such as number of customers of a certain type without service, are imperfect but convey less false precision than the economic valuations based on simple person-days times a constant. They also connect directly to societal "Level of Service" goals that generally accompany major seismic investments in water systems. In essence, a utility might establish a goal of "restore service to critical customers within 7 days and all customers within 60 days" for a given earthquake scenario. This is more tangible than a promise to maintain the number of lost person-days of water to a certain level.

SECTION G.3: VULNERABILITY

Vulnerability is the V term in the J100-21 risk equation $R = C \times V \times T$; it represents the likelihood that an estimated consequence C for a given threat will cause a loss of critical function for a specific asset.

One way to estimate the vulnerability of a water system component to earthquake damage is using fragility estimates, which are equations relating expected loss of function to an earthquake metric such as PGA, PGV, and PGD. For a high-level review of a system, it's often sufficient to use published fragility estimates, rather than facility-specific estimates.

HAZUS MH, Technical Manual Chapter 8 (FEMA, 2020) contains fragility relationships for many water and wastewater facilities including treatment plants, pump stations, and tanks (see example in Figure G.1). HAZUS provides descriptions of four damage states:

- minor: malfunction of the plant for less than 3 days due to loss of power and slight damage to structures;
- moderate: loss of power for a week, considerable damage to mechanical and electrical equipment or moderate damage to structures;
- extensive: structure extensively damaged or pumps damaged beyond repair; and
- complete: structure collapses.

The fragility curves describe only the performance of the asset under consideration, and do not take account of any upstream dependencies such as power or communication systems.

Looking closely at a fragility curve, one sees that the curve is representing not damage state against a metric but *probability* of that

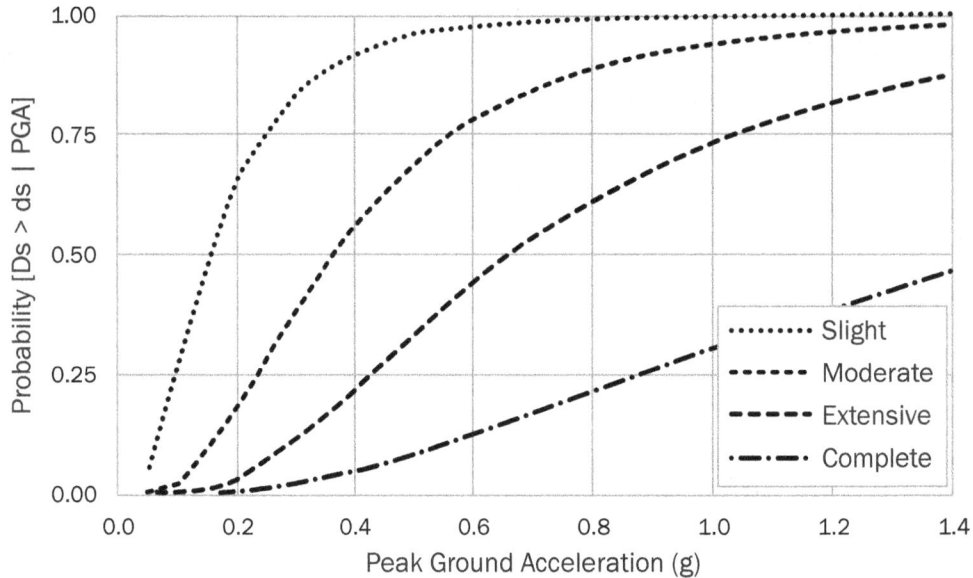

Source: HAZUS Earthquake Model Technical Manual, HAZUS 4.2 SP3, FEMA, October 2020 (Figure 8-11)

Figure G.1   Fragility curves for small pumping plants with anchored components

damage state versus the metric. The natural question then is exactly what stochastic characteristics are being modeled. For example, is the uncertainty associated with the variabilities in the specific earthquake event itself (i.e., an aleatory uncertainty), reflecting that not all earthquakes of a given PGA are equal? Or is it associated with the variabilities in the structures being modeled? If the latter, then in theory a more detailed study of a given facility would reduce that uncertainty and the fragility curves would resemble near-vertical lines. That is, after sufficient study, one would opine that a facility could withstand certain quakes and not others rather than relying on curves that show a wide range of possible outcomes even for a single value of PGA. The HAZUS fragility curves encapsulate all types of uncertainty; this is consistent with their purpose, which is to provide a *rough estimate*, with sparse data, of potential damage to a vast portfolio of assets. This fundamentally low precision should be kept in mind when interpreting results of an analysis.

The fragility curves are typically presented as a family of curves (as shown in Figure G-1) to allow distinctions between damage states.

For purposes of a J100-21 analysis, typically the single curve reflecting moderate damage can be used to simplify the analysis, based on the assumption that minor damage, defined as less than three days of downtime, is tolerable for most assets. For example, if the scenario earthquake being analyzed has a PGA of 0.6g, the vulnerability value for the asset shown in Figure G.1 would be 0.77. There are other valid approaches, but this specific method fits well in the J100-21 framework by providing a value for Vulnerability. It can be adjusted to circumstances; for example, specific assets for which downtime of three days would cause unacceptable consequences could be modeled using the vulnerability associated with minor damage—in this example 0.95.

For important facilities, precision and accuracy can be improved by either adjusting existing fragility curves or creating new curves based on a structural analysis of the facility under consideration. Such an analysis would be focused on the perceived "weak link" of a facility. For example, the user may want to establish whether a steel tank's walls will uplift during the scenario event. Rather than recreating design calculations, reviewing the tank's general design features and its design vintage (i.e., which codes were used for its design) might provide insight into the appropriate fragility.

A final note on the fragility curves is that their mathematical formulation lends itself to direct calculation if desired, instead of visually reading the curves. Tabulations of median, aka PGA_0 and Beta, appear in HAZUS accompanying the various fragility curves. Given a value for PGA at a point of interest, one can compute the probability of damage as:

$$P = \text{phi}[1/\text{Beta} \times \ln(PGA/PGA_0)]$$

where phi is the standard normal cumulative distribution function.

In Excel, the formula NORMDIST can be used, supplying values of 0 for mean and 1 for standard deviation. This formula works for PGA and not necessarily for other metrics, for which the HAZUS technical manual provides the formulations.

SECTION G.4: THREAT LIKELIHOOD

The foregoing discussion describes how to estimate consequences and vulnerability to specific earthquake scenarios. For use in J100-21, the Threat level T is needed as part of the formulation $R = C \times V \times T$; i.e., how likely is the scenario to occur on an annual basis? The value of T is $1/R$, where R is the recurrence interval of the earthquake scenario in years.

As discussed in Sec. G.1, two distinct recurrence intervals are usually considered for earthquakes:

- 1-in-475-year events, equivalent to 10% exceedance in 50 years
- 1-in-2,475-year events, equivalent to 2% exceedance in 50 years

It's often the case that specific "scenario earthquakes," whether historic events or postulated events, will have a difference recurrence interval R. The general rule $T=1/R$ will apply regardless.

Sec. G.4.1 **Basic Screening Approach to Earthquake Risk Analysis**

A first step in earthquake risk analysis is using available online maps to determine whether risks are acceptably low or warrant further analysis. Figure G.2 shows the wide range of seismic risk in the continental United States as an example.

The map shown is based on a 100-year planning horizon, not 50 years as for the derivation of recurrence intervals. To a first approximation, the criterion "<4% in 100 years" is similar to "<2% in 50 years," i.e., the blue areas shown on the map with risk of a damaging earthquake at less than 4% in 100 years are areas for which major earthquakes recur at intervals greater than 2,475 years. Water utility owners in such areas might reasonably conclude their seismic risk analysis at that point, while other owners, i.e., those in areas with clearly greater risk, might proceed with more analytical steps as outlined subsequently.

Sec. G.4.2 **Steps to Take When Screening Shows a Risk**

When further analysis *is* needed, that analysis will begin with identifying one or more credible earthquake scenarios as described in Sec. G.2. Each scenario should be associated with a recurrence R, and thus with an associated $T=1/R$ for use in the J100-21 risk equation.

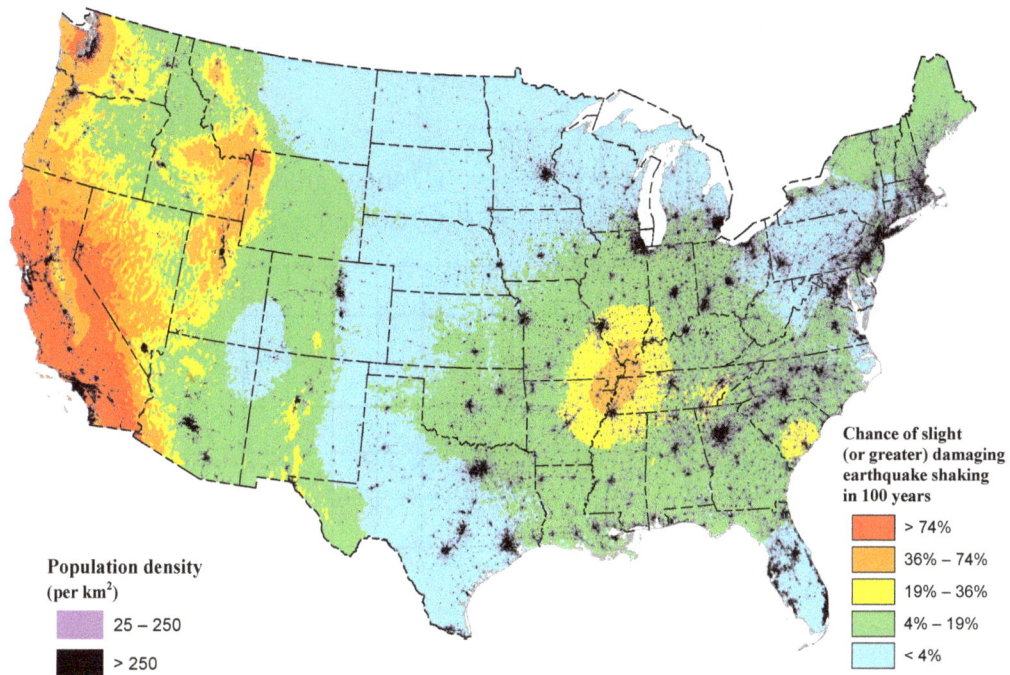

Population density
(per km²)

25 – 250

> 250

Chance of slight
(or greater) damaging
earthquake shaking
in 100 years

> 74%

36% – 74%

19% – 36%

4% – 19%

< 4%

Source: 2018 National Seismic Hazard Map (NSHM)

Figure G.2   National seismic hazard map

SECTION G.5:   EXAMPLE

The example shown in Figure G.3 has been created to show how utilities can estimate the impact on their water system resulting from seismic events. The example generates results for the damage state, the restoration times, and the annual outage risk cost.

The supply of the system is a single well with a capacity of 1,000 GPM. There are two tanks providing storage in the system. There is an older elevated steel storage tank, with a capacity of 100,000 gal, located closer to the river. There is also a newer anchored ground-level concrete storage tank, also with a capacity of 100,000 gal, farther from the supply well and located at higher elevation. There are 30,000 ft of pipe: 15,000 ft cast iron (CIP), 9,000 ft unrestrained joint ductile iron (DIP), and 6,000 ft steel with welded joints.

The town has 500 services and a population of 1,500. The people use 100 gal per person per day. The town uses 150,000 GPD.

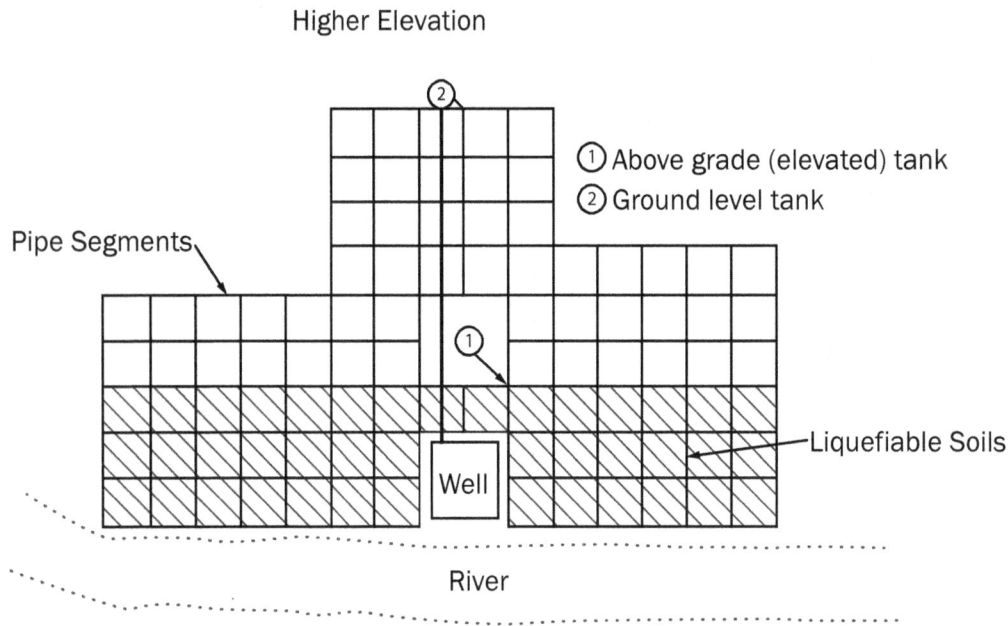

Higher Elevation

① Above grade (elevated) tank
② Ground level tank

Pipe Segments

Liquefiable Soils

Well

River

Figure G.3   Example system for seismic analysis

There are liquefiable soils along the river. Eighty percent of the CIP is in liquefiable soils, and none of the DIP or steel pipe is in liquefiable soils. The area closest to the river developed first.

Sec. G.5.1   **Identify Scenario Earthquakes and Metrics for Each**

There are two seismic events being considered in this example. One is a great earthquake (e.g., a Cascadia Subduction Zone [CSZ] earthquake in the Pacific Northwest) of magnitude 9.0 (M9.0), which occurs 100 mi away from the town, with 3 min of strong ground shaking. The second event is a moderate earthquake of magnitude 6.0 (M6.0), which occurs near the town, with 10 s of strong ground shaking.

To evaluate the seismic impacts, it is necessary to estimate the peak ground acceleration (PGA) and the peak horizontal ground velocity (PGV). In the example, it is assumed that somehow an estimate of the peak ground displacement (PGD) is also available, although that is often not the case. It is also necessary to have an estimate of the percent of liquefaction zone that will liquefy (L_S).

For the seismic events being analyzed in this example, the following values were used:

- M9.0 event: the PGA is 22% of gravity; the PGV is 12 in./s; the PGD is 9 in., and 25% of soil in liquefaction zone liquefies.
- M6.0 event: PGA is 44% of gravity; PGV is 17 in./s; the PGD is 8 in., and 15% of soil in liquefaction zone liquefies.

Note that the M9.0 event produces a smaller shaking intensity than the M6.0 event because it is much further away. However, the percent of area that liquefies and the PGD may be larger for longer-duration earthquakes with the same shaking intensity at a specific location.

Values of PGA and PGV are available in USGS Shakemap scenarios that have been developed for many urban areas across the United States.

Liquefaction susceptibility maps are often available from the state department of geology, particularly in urban areas. HAZUS provides guidance on the percent of area that will liquefy to the extent that lateral spreading will occur (i.e., PGD). If no mapping is available for PGD estimates, the evaluator can use a value of 6 in., a moderate level of PGD, in areas mapped as being at significant risk of liquefaction and zero elsewhere.

Sec. G.5.2   **Estimate Consequence and Vulnerability**

This section uses the HAZUS MH Technical Manual (FEMA 2020) as the basis for establishing the damage state and restoration times for the example system facilities, the well, and two tanks. The referenced tables are from the HAZUS manual and provide a median of the damage state. Fragility curves were described in Sec. G.3. The evaluator may choose to perform a more detailed assessment and examine probabilities of multiple damage states, but in the basic case, the assessment is performed for moderate damage, since lesser damage levels by definition result in outages of less than three days and *should* have minimal societal impact.

Damage to Well

The vulnerabilities for the well comes from Figure 8-15 or are alternatively computed using Table 8-9 (both in the HAZUS MH technical manual) and are based on the PGA value of 0.44g and 0.22g for the two scenarios; the probabilities corresponding to moderate damage are 0.62 and 0.22.

Assume that the well may cost $3,000 to repair if damaged (C), so C × V is:

- M6 event: $3,000 × 0.62 = $1,860
- M9 event: $3,000 × 0.62 = $660

Damage to Tanks

The damage states for tanks come from Figures 8-20 and 8-21 or are alternatively computed using the values in Table 8-10 (both in the HAZUS MH technical manual) and are also based on the PGA.

For the above ground steel storage tank (age is not a factor in the evaluation in HAZUS), the probabilities of moderate damage are 0.33 and 0.03 for the M6.0 and M9.0 events, respectively.

For the anchored on-ground concrete storage tank, the probabilities of moderate damage are 0.41 and 0.11 for the M6.0 and M9.0 earthquakes, respectively.

There is considerable uncertainty about the tank repair cost. Assume $4,000 in repair costs, for each damaged tank, bearing in mind that the vulnerabilities (i.e., the likelihood of the tanks being moderately damaged), varied by tank and by scenario. Thus, the average C × V values for tank repairs are as follows:

- Tank repair for M6 event = $4,000 × [.33+.41] = $2,960
- Tank repair for M9 event = $4,000 × [.03+.11] = $560

Post-earthquake Facility Restoration Time

The description of damage and the associated restoration times are described in Chapter 8 of HAZUS MH 4.2 Technical Manual. For establishing the actual restoration times, use Tables 8-1 and 8-2 and Figures 8-3 and 8-4. These tables provide the percent likelihood of full restoration by the given day. For example, for a well with slight/minor damage, there is an 85 percent chance the well will be restored in 1 day; for a well with moderate damage, there is a 90 percent chance it will be restored within 3 days. To be consistent with the use of moderate damage to estimate vulnerability V, we use the same damage state for restoration time, or seven days per Table 8.1.b.

Pipe Damage

This section uses the ALA Seismic Fragility Formulations for Water Systems (American Lifelines Alliance, 2001). Only pipe damage is considered here. Service line (utility and customer side) breaks and leaks are not included.

Pipe Damage from PGV

The PGV is 17 in./s and 12 in./s for the magnitude 6 and 9 events, respectively. The repair rate (RR) due to PGV is:

Repair rate/1,000 ft = $K_1 \times 0.00187 \times PGV$

Cast Iron

The value of K_1 for cast iron from Table G.1 is 1.0.

- RR = 1.0 × .00187 × 17 in./s, for the M6 event = 0.032 repairs/1,000 ft of pipe
- RR = 1.0 × .00187 × 12 in./s, for the M9 event = 0.022 repairs/1,000 ft of pipe

There are 15,000 ft of cast iron pipe in this example system.

The number of cast iron pipe repairs, for the M6 event = 15 × 0.032 = 0.48 repairs

The number of cast iron pipe repairs, for the M9.0 event = 15 × 0.022 = 0.33 repairs

Ductile Iron

The value of K_1 for ductile iron from Table G.1 is 0.5.

- RR = 0.5 × .00187 × 17 in./s, for the M6.0 event = 0.016 repairs/1,000 ft of pipe
- RR = 0.5 × .00187 × 12 in./s, for the M9.0 event = 0.011 repairs/1,000 ft of pipe

There is 9,000 ft of ductile iron pipe in this example system.

The number of ductile iron pipe repairs, for the M6.0 event = 9 × 0.017 = 0.15 repairs

The number of ductile iron pipe repairs, for the M9.0 event = 9 × 0.011 = 0.1 repairs

Steel

The value of K_1 for steel from Table G.1 is 0.60 (an average value for small welded joint steel).

- RR = 0.60 × .00187 × 17 in./s, for the M6 event = 0.019 repairs/1,000 ft of pipe
- RR = 0.60 × .00187 × 12 in./s, for the M9.0 event = 0.0135 repairs/1,000 ft of pipe

There is 6,000 ft of steel pipe in this example system.

The number of steel pipe repairs, for the M6.0 event = 6 × .019 = 0.11 repairs

The number of steel pipe repairs, for the M9.0 event = 6 × 0.0135 = 0.08 repairs

Sum up the total number of expected PGV-related repairs for the three pipe materials for each earthquake:

- The number of repairs from PGV for the M6.0 event = 0.48 + 0.15 + 0.11 = 0.74 (use 0.7) repairs
- The number of repairs from PGV for the M9.0 event = 0.33 + 0.1 + 0.08 = 0.51 (use 0.5) repairs (interpret this to mean there will be between 0 and 1 repair only due to PGV).

As these failures are due to PGV, 80% are expected to be leaks and 20% breaks:

- M6.0 event: 0.7 repairs × 0.8 = 0.1 breaks and 0.6 leaks
- M9.0 event: 0.5 repairs × 0.8 = 0.1 breaks and 0.4 leaks

Pipe Damage from PGD

The peak ground deformation (PGD) is 8 in. and 9 in., respectively, for the M6.0 and M9.0 events. Also, 15% and 25% of soil in liquefaction zone liquefies, respectively, for the M6.0 and M9.0 events.

The repair rate (RR) from PGD is:

Repair rate/1,000 ft = $K_2 \times 1.06 \times PGD^{0.319} \times L_S$

Cast Iron

The value of K_2 for cast iron from Table G.1 is 1.0.

- RR = $1.0 \times 1.06 \times (8 \text{ in./s})^{0.319} \times 0.15$ for the M6.0 event = 0.31 repairs/1,000 ft of pipe
- RR = $1.0 \times 1.06 \times (9 \text{ in./s})^{0.319} \times 0.25$ for the M9 event = 0.53 repairs/1,000 ft of pipe

There is 15,000 ft of cast iron pipe in this example system; 80% of that, or 12,000 ft of cast iron pipe, is in liquefiable soils.

204

Table G.2 Summary of example pipeline repairs

Event	PGV Leaks	PGV Breaks	PGD Leaks	PGD Breaks	Total Leaks	Total Breaks
M6.0	0.6	0.1	0.7	3.0	1.3 (1)	3.1 (3)
M9.0	0.4	0.1	1.3	5.1	1.7 (2)	5.2 (5)

The number of cast iron pipe repairs due to PGD, for the M6.0 event = 12 × 0.31 = 3.7 repairs.

The number of cast iron pipe repairs due to PGD, for the M9.0 event = 12 × 0.53 = 6.4 repairs.

There is no ductile iron or steel pipe in liquefiable areas.

As these failures are due to PGD, 20% are expected to be leaks and 80% breaks:

- M6.0 event: 3.7 repairs × 0.8 = 3.0 breaks and 0.7 leaks
- M9.0 event: 6.4 repairs × 0.8 = 5.1 breaks and 1.3 leaks

The total of leaks and breaks for each event is shown in Table G.2:

The cost for leak repair is set at $2,000 (based on a crew spending 4 h to complete the work). The cost for a break repair is set at $5,000 (based on a full day of work plus material cost). Thus, the cost of pipe repair for the two scenarios is $17,000 for M6 and $29,000 for M9.

Post-earthquake Pipeline Restoration Time

Assume in this example that the utility's one crew is available, and they can do two leak repairs per day and one break repair per day (from HAZUS, pages 8–11 and 8–12). This is highly dependent on pipe diameter, crew experience, the type of equipment available, and whether repair materials are available.

For the M6.0 event, there is one leak repair and three break repairs.

For the M9.0 event, there are two leak repairs and five break repairs.

It would take 4 days for the M6.0 event and 6 days for M9.0 event (5 days to repair the breaks and 1 day for the leaks).

The facility and pipeline restoration time is summarized in Table G.3.

For this example, assume that all the pipelines must be repaired before any of the system is operable and that one of the tanks must be operable before any of the system is operable. Depending on the system design, it may be possible to operate the system after some of the pipelines are repaired and/or with no tanks. The tanks will not necessarily be damaged (i.e., their V value was low, reflecting low likelihood of moderate or worse damage). A more refined analysis could examine system restoration times stochastically.

For the M9.0, the earthquake ground motions are moderate as the event is distant, but the duration results in a higher rate of liquefaction and associated PGD and pipeline damage. It would take an estimated 6 days for all of the pipelines to be repaired. The M6.0 event has higher ground motions resulting in more damage to facilities, especially the tanks. Liquefaction, however, is limited because of its shorter duration.

Societal Consequences

Because in either scenario the repairs last for 6–7 days, assume utilities cannot get water to customers during that time (it may be possible to pump directly from the well to the customer, but assume it is not). The outage cost associated with a loss of supply is $105/person-day (FEMA Benefit–Cost Analysis Sustainment and Enhancements, 2022). Therefore, the societal outage cost for either event = 1,500 population x $116/day/person × 7 days = $1,218,000.

Table G.3 Facility and pipeline restoration time

Event	Well	Aboveground Tank	Ground-Level Tank	Pipelines	Maximum
M6.0	3	7	7	4	7
M9.0	1	1	1	6	6

Table G.4 Total Consequences × Vulnerabilities

Scenario	M6	M9
Well	$1,860	$660
Tanks	$2,960	$560
Pipes	$17,000	$29,000
Subtotal direct consequence	$21,820	$28,220
Societal consequence	$1,218,000	$1,218,000
Total consequence	$1,261,640	$1,248,220

Sec. G.5.3 Threat Likelihood

In this analysis, we've considered two scenario earthquakes: an M6.0 and an M9.9 earthquake, with recurrence intervals of 100 and 500 years. These values correspond to values of T in the J100-21 risk equations of 1/100 and 1/500, i.e., 0.01 and 0.002 as annual probabilities.

Sec. G.5.4 Risk Quantification

The final step in this analysis is to estimate the risk of each scenario using the standard J100-21 formulation R = C × V × T "risk cost" for each considered scenario. In practice, since there is a mix of assets with varying vulnerabilities, this equation could be considered as the sum of C × V for the various assets, and societal consequences, times the T value.

- R = \$1,261,640 × 0.01 = \$12,616 for M6
- R = \$1,248,220 × 0.002 = \$2,496 for M9

Clearly, there is a much higher risk cost associated with the M6.0 event. Computing the aggregate risk posed by a suite of scenario earthquakes is nontrivial for several reasons. It's recommended for the current purpose that a utility examine several scenarios and then pick a limited number for economic valuation purposes.[¶]

The annualized losses from the individual events can be used to develop benefit–cost analyses to be used in the FEMA Hazard Mitigation Program applications and for justification and/or prioritization of capital improvement projects.

SECTION G.6:   REFERENCES

American Lifelines Alliance (ALA). 2001. "Seismic Fragility Formulations for Water Systems." Washington, DC: ALA.

American Lifelines Alliance. 2005. "Guidelines for Implementing Performance Assessments of Water Systems." Washington, DC: ALA.

[¶] As just a hint of the challenges, consider that there are an infinite number of scenarios possible, each with its own risk. Yet intuitively we know that total risk is not infinite, as we would expect if we simply summed the scenarios. We know this because in fact the occurrence of a quake at a specific time and place greatly impacts the conditional probability of other earthquakes in the region. That's because earthquakes are not simple Poisson processes.

AWWA. 2021. J100. Risk and Resilience Management of Water and Wastewater Systems. Denver: AWWA.

FEMA. 2020. *HAZUS MH 4.2 Earthquake Model Technical Manual.* Washington, DC: FEMA.

Petersen, M.D., A.M. Shumway, P.M. Powers, C.S. Mueller, M.P. Moschetti, et al. 2019. "The 2018 Update of the US National Seismic Hazard Model: Overview of Model and Implications." *Earthquake Spectra.* 36(1):5.

APPENDIX H

Tornado

SECTION H.1: INTRODUCTION

A tornado consists of a violently rotating column of air varying in size and intensity. While poorly understood, they are typically spawned by supercells (rotating thunderstorms with well-defined radial circulation) or by hurricanes and, by definition, must be in contact with both the ground and the base of the clouds. Supercells and hurricanes are often observed as creating several tornadoes in a single storm event. The characteristic "funnel cloud" may or may not be visible. Tornadoes can involve wind speeds higher than any other natural hazard, but measurement of "true" wind speed at ground level inside a tornado is virtually impossible (www.spc.noaa.gov/faq/tornado/#Damage).

Damages from tornadoes are caused by two primary factors: direct wind pressure and a vacuum effect that forms in the center of the storm. Flying debris can cause additional damage. Hail, rain, and lightning may accompany the high winds. The high-velocity air in the center of the funnel produces a partial vacuum inside the tornado. The local pressure inside the funnel is quite low compared to normal atmospheric pressure because of the extremely high winds in a tornado and the small diameter.

Tornadoes are local phenomena that move at a relatively high velocity. Tornadoes quickly reduce the external pressure around an object without allowing time for the internal pressure to equalize with the lowered external pressure. A closed structure, such as a house, will literally explode when the tornado passes over it. The higher internal pressure inside the house will cause the walls and roof to explode outward, destroying the integrity of the structure. The high-velocity winds can then demolish the remaining structure.

Due to difficulties in measuring the internal wind speed of a tornado as well as fluctuating wind speeds along its track, tornado intensity is determined by observable damage levels, with expert judgment used to estimate specific wind-speed ranges. The most

Table H.1 Tornado scale and typical damage

Enhanced Fujita Scale		
EF Number	**3-Second Gust (mph)**	**Typical Damage**
EF0	65–85	**Light:** Some damage to chimneys; branches broken off trees; shallow-rooted trees pushed over; sign boards damaged.
EF1	86–110	**Moderate:** Peels surface off roofs; mobile homes pushed off foundations or overturned; moving autos blown off roads.
EF2	111–135	**Considerable:** Roofs torn off frame houses; mobile homes demolished; boxcars overturned; large trees snapped or uprooted; light-object missiles generated; cars lifted off ground.
EF3	136–165	**Severe:** Roofs and some walls torn off well-constructed houses; trains overturned; most trees in forest uprooted; heavy cars lifted off the ground and thrown.
EF4	166–200	**Devastating:** Well-constructed houses leveled; structures with weak foundations blown away some distance; cars thrown and large missiles generated.
EF5	>200	**Incredible:** Strong frame houses leveled off foundations and swept away; automobile-sized missiles fly through the air in excess of 100 m (109 yds); trees debarked; incredible phenomena will occur.

Source: Storm Prediction Center, National Oceanic and Atmospheric Administration (NOAA)

common measure of tornado intensity used in the United States and Canada is the Enhanced Fujita Scale, or EF-Scale, which was updated from the earlier Fujita Scale (or F-Scale) in 2007. The J100-21 reference threats T0 through T5 correspond to the Fujita scales of EF0 through EF5. Table H.1 shows the estimated 3-second wind gust and typical damage for the six EF-scale categories.

Two characteristics are relevant to analyzing tornado risk at a specific geographic location. The vast majority of tornadoes are those with lower wind speeds—almost 80% are EF0 or EF1 and do only a modest degree of damage. However, there is a very strong positive correlation between the EF class and the area contacted; an EF4 tornado affects an area more than 200 times on average than that affected by an EF0 tornado. Because of the widely divergent frequencies and areas, it is necessary to analyze each class, respectively.

SECTION H.2: CONSEQUENCES

The preceding discussion explains why certain types of structures are more likely to be demolished by a tornado than others. Open space-frame type structures, such as piping and slab-mounted equipment, pipe racks, beam and column frames, freestanding pressure vessels, and machinery, will be affected by the high-velocity winds, but the pressure differential does not typically cause damage. Closed structures are much more likely to be demolished. However, blast-resistant structures, such as control rooms for refineries, underground storage for water treatment facilities, and bunkers used for storing explosives and military equipment, etc., have the capability to survive tornadoes.

For the purposes of this analysis, it is assumed that damage due to any category or magnitude tornado could cause damage to buildings and equipment. The damage factors for tornadoes are provided in Table H.2. For estimating tornado-loss consequence, the repair or replacement cost is multiplied by the damage factor of the asset from the table. Anticipated lost revenue, loss of life, and serious injuries should be added to this value. As an alternative to these damage factors, best engineering judgment can be used to estimate physical damage.

Table H.2 Tornado damage factors

Tornado Damage Factors	Asset Types and Mountings
EF 0–1 0.01 EF 2–3 0.25 EF 4–5 1.0	Slab-mounted equipment—pumps, valves, compressors, meters, electric motors, electrical controls, consoles, etc. Ground-level storage tanks
EF 0–1 0.05 EF 2–3 0.5 EF 4–5 1.0	Above ground piping designed to accepted codes and standards such as ASME B31.1, ASME B31.3 Pressure vessels designed to ASME codes and standards Elevated storage tanks Buildings designed to UBC Code or equivalent
EF 0–1 0.1 EF 2–3 1.0 EF 4–5 1.0	Buildings not designed to code Portable buildings and trailers Automobiles and trucks, heavy equipment

SECTION H.3:   VULNERABILITY

The vulnerability to EF0 through EF5 tornadoes is assumed to be 1.0 in consistency with J100-10.

SECTION H.4:   THREAT LIKELIHOOD

Reliable data exist to determine the threat likelihood of tornadoes across the United States. The frequency of tornadoes is based on the average annual number (N) of tornadoes in a given county or state multiplied by the ratio of the average area for a single tornado (TA) divided by the total area (A) of the county or state. In equation form:

Threat Likelihood = N × (TA/A)

Where:

N = average annual number of tornadoes in the county or state (see Table H.4 for state values)

TA = average area of a tornado (mi^2) (see Table H.3)

A   = area of county or state (mi^2) (see Table H.4 for state values)

Note that if N is provided at a state level, then A must be the area of the state. If N is provided at the county level, then A must be the area of the county. Use of the smaller geographic unit generally improves the estimated losses by reducing the amount of variability in key factors.

In the United States, the average tornado has a 3.62-mi length, 0.062-mi width, and 0.61-mi^2 area. The average area of a tornado (TA) for each magnitude of tornado is shown in Table H.3.

Table H.4 provides the average annual number of tornadoes (N) by state (and the District of Columbia) for each magnitude of tornado and the area of each state. Raw data was provided by NOAA for 1950–2019 and can be found at https://www.spc.noaa.gov/wcm/#data. If there are large variations of tornado occurrences across a state or if more detailed frequencies are preferred for the analysis, the raw database contains data down to the county level identified by Federal Information Processing System (FIPS) code. If county data is used,

Table H.3 Average area by tornado magnitude

Tornado Magnitude	Average Area of Tornado (TA) (mi^2)
EF0	0.051
EF1	0.324
EF2	1.124
EF3	4.208
EF4	10.523
EF5	18.679

then the area of the county should be used in the equation instead of the area of the state.

Examples

Asset Type: Elevated storage tank

Location (State): Alabama

Replacement Cost: $1,000,000

EF0 Tornado

Consequences

C(0):

Repair or Replacement = $1,000,000 × 0.05 = $50,000

Service Loss: $0

C(0) = $50,000 + $0

C(0) = $50,000

Vulnerability

V(0) = 1.0

Threat Likelihood

T(0) = N(0) × (TA(0)/A)

T(0) = 10.857 × (0.051/52,419)

T(0) = 0.00001056

Risk

R(0) = C(0) × V(0) × T(0)

R(0) = $50,000 × 1.0 × 0.00001056

R(0) = $1

Table H.4 Average annual number of tornadoes (N) by magnitude and area of state (A)

State	EF0	EF1	EF2	EF3	EF4	EF5	Area of State (mi²)
AK	0.057	0.0	0.0	0.0	0.0	0.0	663,264
AL	10.857	12.643	6.543	2.086	0.571	0.171	52,419
AR	7.629	10.8	6.214	2.314	0.414	0.0	53,179
AZ	2.457	1.029	0.229	0.043	0.0	0.0	113,998
CA	4.343	1.729	0.329	0.029	0.0	0.0	163,695
CO	21.0	8.386	1.729	0.329	0.014	0.0	104,184
CT	0.329	0.829	0.314	0.057	0.029	0.0	5,543
DC	0.057	0.0	0.0	0.0	0.0	0.0	68
DE	0.343	0.386	0.186	0.014	0.0	0.0	2,491
FL	29.571	13.929	4.671	0.586	0.043	0.0	65,754
GA	7.3	11.671	4.729	1.114	0.157	0.0	59,424
HI	0.4	0.129	0.057	0.0	0.0	0.0	10,931
IA	16.671	12.1	6.486	1.729	0.7	0.1	56,271
ID	2.1	0.871	0.143	0.0	0.0	0.0	83,642
IL	16.843	12.2	5.943	1.686	0.486	0.029	57,914
IN	7.2	8.571	4.414	1.514	0.543	0.071	36,417
KS	36.343	15.471	6.671	2.957	0.686	0.129	82,276
KY	4.157	6.4	3.071	1.171	0.314	0.043	40,409
LA	9.7	14.814	4.943	1.457	0.129	0.029	52,271
MA	0.571	1.314	0.514	0.086	0.043	0.0	10,554
MD	2.357	2.414	0.457	0.1	0.029	0.0	12,407
ME	0.486	1.171	0.271	0.0	0.0	0.0	35,385
MI	5.2	6.029	3.057	0.6	0.243	0.029	96,715
MN	14.743	8.629	2.986	0.814	0.343	0.043	86,942
MO	13.271	13.929	5.229	1.671	0.7	0.029	69,704
MS	9.971	14.586	6.586	2.286	0.543	0.129	48,434
MT	3.957	1.371	0.629	0.157	0.014	0.0	147,164
NC	8.514	7.629	3.0	0.543	0.186	0.0	53,819
ND	14.371	5.686	1.929	0.557	0.171	0.043	70,761
NE	22.614	12.7	4.571	1.443	0.529	0.029	77,421
NH	0.4	0.614	0.314	0.057	0.0	0.0	9,350
NJ	0.886	0.943	0.371	0.086	0.0	0.0	8,729
NM	6.343	1.771	0.586	0.057	0.0	0.0	121,697
NV	1.114	0.229	0.0	0.0	0.0	0.0	110,566
NY	2.443	2.9	0.857	0.3	0.057	0.0	54,556
OH	5.771	6.957	2.9	0.743	0.329	0.071	44,825
OK	23.7	19.157	10.2	2.986	0.914	0.143	69,959

(continued)

Table H.4 Average annual number of tornadoes (N) by magnitude and area of state (A) (*Continued*)

State	EF0	EF1	EF2	EF3	EF4	EF5	Area of State (mi²)
OR	1.314	0.329	0.057	0.029	0.0	0.0	98,466
PA	3.357	6.143	2.371	0.386	0.143	0.014	46,055
RI	0.086	0.114	0.014	0.0	0.0	0.0	1,212
SC	6.214	6.043	2.143	0.443	0.2	0.0	32,020
SD	16.071	5.557	3.486	0.9	0.129	0.014	77,183
TN	5.514	7.271	3.586	1.357	0.471	0.014	42,180
TX	68.2	36.4	17.671	4.743	0.814	0.086	268,819
UT	1.3	0.386	0.129	0.014	0.0	0.0	84,898
VA	4.5	4.686	1.386	0.486	0.029	0.0	42,774
VT	0.171	0.286	0.214	0.014	0.0	0.0	9,623
WA	0.986	0.543	0.186	0.043	0.0	0.0	71,362
WI	7.357	8.1	3.843	0.829	0.271	0.043	65,556
WV	0.686	1.071	0.329	0.114	0.0	0.0	24,230
WY	6.186	2.743	0.857	0.214	0.014	0.0	97,818

Source: Raw data from NOAA

EF1 Tornado

Consequences

C(1):

Repair or Replacement = $1,000,000 × 0.05 = $50,000

Service Loss: $0

C(1) = $50,000 + $0

C(1) = $50,000

Vulnerability

V(1) = 1.0

Threat Likelihood

T(1) = N(1) × (TA(1)/A)

T(1) = 12.643 × (0.324/52,419)

T(1) = 0.0000781

Risk

R(1) = C(1) × V(1) × T(1)

R(1) = $50,000 × 1.0 × 0.0000781

R(1) = $4

EF2 Tornado

Consequences

C(2):

Repair or Replacement = $1,000,000 × 0.5 = $500,000

Service Loss: $5,000

C(2) = $500,000 + $5,000

C(2) = $505,000

Vulnerability

V(2) = 1.0

Threat Likelihood

T(2) = N(0) × (TA(2)/A)

T(2) = 6.543 × (1.124/52,419)

T(2) = 0.0001403

Risk

R(2) = C(2) × V(2) × T(2)

R(2) = $505,000 × 1.0 × 0.0001403

R(2) = $71

EF3 Tornado

Consequences

C(3):

Repair or Replacement = $1,000,000 × 0.5 = $500,000

Service Loss: $7,500

C(3) = $500,000 + $7,500

C(3) = $507,500

Vulnerability

V(3) = 1.0

Threat Likelihood

T(3) = N(3) × (TA(3)/A)

T(3) = 2.086 × (4.208/52,419)

T(3) = 0.00016746

Risk

R(3) = C(3) × V(3) × T(3)

R(3) = $507,500 × 1.0 × 0.00016746

R(3) = $85

EF4 Tornado

Consequences

C(4):

Repair or Replacement = $1,000,000 × 1.0 = $1,000,000

Service Loss: $15,000

C(4) = $1,000,000 + $15,000

C(4) = $1,015,000

Vulnerability

V(4) = 1.0

Threat Likelihood

T(4) = N(4) × (TA(4)/A)

T(4) = 0.571 × (10.523/52,419)

T(4) = 0.00011463

Risk

R(4) = C(4) × V(4) × T(4)

R(4) = $1,015,000 × 1.0 × 0.00011463

R(4) = $116

EF5 Tornado

Consequences

C(5):

Repair or Replacement = $1,000,000 × 1.0 = $1,000,000

Service Loss: $20,000

C(5) = $1,000,000 + $20,000

C(5) = $1,020,000

Vulnerability

V(5) = 1.0

Threat Likelihood

T(5) = N(5) × (TA(5)/A)

T(5) = 0.171 × (18.679/52,419)

T(5) = 0.00006093

Risk

$R(5) = C(5) \times V(5) \times T(5)$

$R(5) = \$1{,}020{,}000 \times 1.0 \times 0.00006093$

$R(5) = \$62$

Total Risk

$\text{Risk} = R(0) + R(1) + R(2) + R(3) + R(4) + R(5)$

$\text{Risk} = \$1 + \$4 + \$71 + \$85 + \$116 + \$62$

Risk = \$339

APPENDIX I

Wildfire

SECTION I.1: INTRODUCTION

Wildfires are common in parts of the country, especially west of the Rocky Mountains. In many cases, the immediate effect of these fires on the water sector is minimal as the typical wildfire is a fast-moving conflagration that sweeps across the system with little direct damage. The lack of trees on the plant site denies the fire a burn path, and the types of structures on a facility's property tend to be only superficially damaged by these types of fires. However, the long-term effects from these fires can be serious for the utility owner/operator. Contamination of the watershed can cause intake clogging, treatment problems, and finished water quality issues.

The approach advocated here to analyze the risk to wildfires uses two separate categories: minor wildfires (J100-21 "W1") and major wildfires (J100-21 "W2"). Minor wildfires are those characterized as consuming an area between 0.01 and 10 km², while major wildfires will be those consuming an area greater than 10 km². This is similar to the graduated scales of hurricanes and earthquakes and will require two separate sets of calculations.

SECTION I.2: CONSEQUENCES

The consequences to the facility owner from a wildfire are the sum of the individual losses. The potential financial consequences could include many variables. The following are the most common issues that should be considered when calculating the consequences from a wildfire.

Cost of asset repair/replacement. This is the dollar value of the asset if it needs to be repaired or replaced.

Service outage due to the time required to repair/replace the asset. This is the dollar value of income lost (time out of service ×

MGD lost × $/MGD) because the asset is unusable due to needed repairs/replacement. This includes the time to order, access, and repair/replace the asset.

Service outage due to power loss. This is the dollar value of income lost (time out of service × MGD lost × $/MGD) because the asset is unusable due to a loss of commercial and any on-site backup power.

Cost of additional treatment needed over the next year. This value describes the additional cost of chemicals, filters, etc., that would be needed through the year of operation after the wildfire to continue to provide water at current quality standards.

Lost income due to lower productivity from additional treatment. If the wildfire is severe enough to require significant amounts of additional treatment time, the treatment capacity of the plant could be decreased. This could cause water shortages in the served area. The lost income due to this decreased capacity and ability to serve customers is accounted for in this value.

Potential fatalities and serious Injuries. If the wildfire will result in any water-related fatalities or serious injuries, they should be accounted for. Be sure to convert the potential losses to dollars if using the Value of Statistical Life and Injuries in the assessment.

Estimated economic losses to the regional economy. If there is determined to be an economic impact on the community, those costs can also be included.

SECTION I.3: VULNERABILITY

This method for calculating vulnerability of a water utility to wildfires is broken down into several categories. These include power loss, asset loss, water quality/treatment issues, and system pressure losses. To capture the significance of each of these consequences, a group of six questions was created. These questions evaluate the vulnerability of the asset to each area of interest, and then a final vulnerability value is derived based on the answers to those questions. There are three potential answers to each of the questions as shown in Table I.1. Other methods may be used to calculate vulnerability.

Table I.1 Value scale for wildfire impact questions

Value	Meaning
0	No impact on the asset
0.5	Moderate impact on the asset
1	High impact on the asset

The questions used to determine the vulnerability of an asset are:

1. Will the asset lose power?
   - 0: If the asset has no threat of losing commercial power because it uses no electricity or has sufficient backup generation on-site to last until power is restored.
   - 0.5: If the asset could/will lose commercial power and there is some backup generation on-site to last a short amount of time but would not be sufficient if commercial power is not restored quickly.
   - 1.0: If the asset will lose commercial power and there is no backup generation on-site. The asset will remain inoperable until commercial power is restored.

2. Will access to the asset be lost?
   - 0.0: There are multiple routes to the asset that would not be affected by a wildfire.
   - 0.5: There are multiple routes to the asset, but they may be affected by the fires, rendering them impassable.
   - 1.0: There is only one route to the asset, and it could be closed due to a wildfire.

3. Will the asset need to be replaced?
   - 0: The asset is buried (underground piping), comprised entirely of materials that do not burn, or has been completely safeguarded against a fire through other protection measures.
   - 0.5: The asset would incur some damage from the fire but would not need complete replacement.
   - 1.0: The asset would need to be replaced due to extensive damage from the fire.

4. Will the source water be lost?
- 0.0: If there is no significant change in the quality or quantity of the source water in the first year following the fire.
- 0.5: If either the quantity or quality of the source water requires a significant increase in treatment time resulting in the inability of the plant to meet its peak demand.
- 1.0: If either the quantity or quality of the source water will not permit the plant to meet its rated average summer-day production.

5. Will the system demand to fight the fire cause the utility to lose system pressure?
- 0.0: If the system is generally able to meet anticipated increased fire flow demands without endangering other users on the system.
- 0.5: If the additional demand will adversely impact some of the service area but not all.
- 1.0: If the additional water used will cause system pressures to drop below state minimum pressure standards (typically 20 psi).

6. Will the wildfire result in significantly increased operating cost for this asset?
- 0.0: If there is anticipated to be only minor additional long-term operating costs because of the fire.
- 0.5: If there is anticipated to be some additional treatment costs, but none that would severely impact the treatment process or output of the facility.
- 1.0: If a significant increase in treatment cost is anticipated with daily treatment capacity limited or the facility can no longer operate.

Once these questions are evaluated, the average of the values determines the overall vulnerability of the asset to wildfires. If a question is not applicable to the asset type, eliminate the question from the calculation. (If only five questions are applicable, take the average of the five questions answered.)

SECTION I.4:   THREAT LIKELIHOOD

The US Forest Service uses Bailey's ecoregion divisions as an accurate depiction of the ecoregions in the United States. These ecoregions are based on certain climate, vegetation, and soil characteristics common to the areas. Figure I.1 depicts Bailey's ecoregions. The regions with a prefix of "M" are noted as mountainous regions.

Table I.2 shows each of these regions, their areas, and frequency of wildfire occurrence.

The threat likelihood is the probability that a wildfire will occur within the facility's watershed in a timeframe of one year. The entire watershed is considered the affected area due to water quality (WQ) issues that will occur at the treatment plant due to the fire occurring elsewhere in the watershed. Threat likelihood for a wildfire is therefore given as:

$$T = \frac{A_w}{A_E} \times \frac{1}{t_E}$$

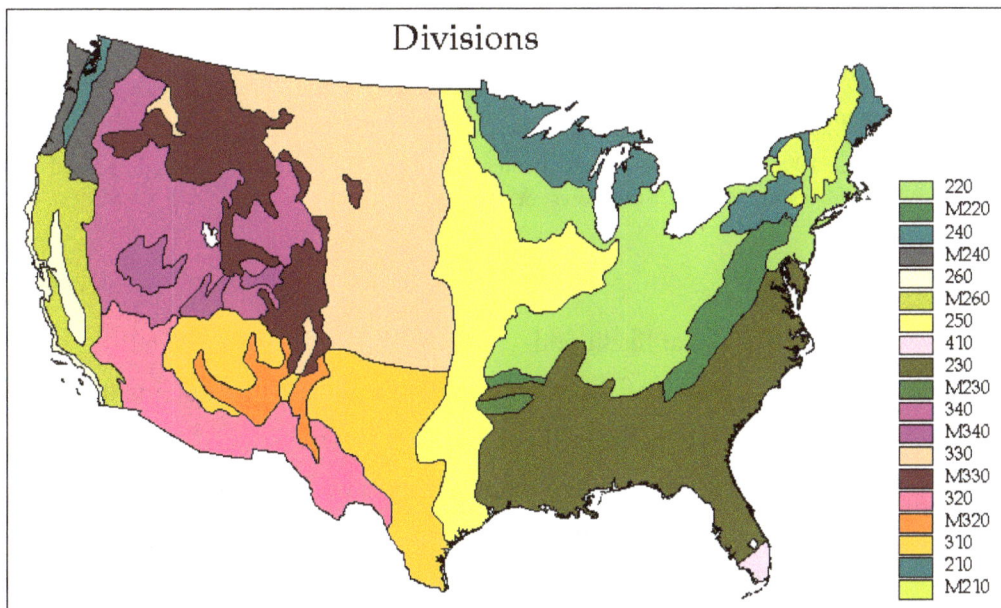

Source: US Forest Service

Figure I.1   Ecoregions in the United States

223

Table I.2   Ecoregion areas and wildfire occurrence

Ecoregion Division Name	Ecoregion Division Code	Ecoregion Division Area, km² (A$_E$)	Minor WF Recurrence Interval, years (t$_E$ Minor)	Major WF Recurrence Interval, years (t$_E$ Major)
Hot Continental	220	969,955	0.19	34
Hot Continental Mountains	M220	192,955	0.30	55
Marine	240	38,591	0.79	23
Marine Mountains	M240	138,300	2.47	100
Mediterranean	260	88,319	0.22	2
Mediterranean Mountains	M260	241,388	0.51	13
Prairie	250	772,597	0.28	9
Savanna	410	20202	0	0
Subtropical	230	1,064,749	0.12	33
Subtropical Mountains	M230	22,792	0.24	32
Temperate Desert	340	689,458	0.85	13
Temperate Desert Mountains	M340	112,924	1.66	27
Temperate Steppe	330	1,099,973	0.89	22
Temperate Steppe Mountains	M330	585,081	1.18	36
Tropical/Subtropical Desert	320	441,811	0.29	8
Tropical/Subtropical Steppe Mountains	M320	130,018	0.41	18
Tropical/Subtropical Steppe	310	654,341	0.40	20
Warm Continental	210	381,507	1.70	203
Warm Continental Mountains	M210	112,924	13.64	1,672

Source: Bailey, R.G. 1995. "Description of the Ecoregions of the United States (2nd ed., revised and enlarged). Misc. Publication 1391. USDA Forest Service.

Where:

T = threat likelihood

A$_W$ = area of the watershed where the facility obtains its source water and where the facility's assets are located, in km²

A$_E$ = ecoregion division area (found in Table I.1), in km²

t$_E$ = wildfire recurrence intervals (found in Table I.1), in years.

This calculation will be completed twice, once for minor wildfires and once for major wildfires, with t$_E$ being the only difference between the two. If the utility has local data regarding the probabilities of wildfires, those probabilities should be used instead of the probabilities generated by the proposed methodology above.

SECTION I.5:   TOTAL RISK

The total annual risk (R) of an asset that is affected by a wildfire is the product of the consequences (C_{Mi}), vulnerability (V_{Mi}), and threat likelihood (T_{Mi}) for minor wildfires plus the product of the consequences (C_{Ma}), vulnerability (V_{Ma}), and threat likelihood (T_{Ma}) for major wildfires. The equation depicting this is:

$$R = (C_{Mi} \times V_{Mi} \times T_{Mi}) + (C_{Ma} \times V_{Ma} \times T_{Ma})$$

SECTION I.6:   EXAMPLE

Asset: Booster pump station

Location: Denver, Colo.

Ecoregion: M330—Temperate Steppe Mountains

Area of the watershed where the facility obtains its source water: (A_w) = 90,000 km^2

Minor Wildfire

Consequences

Cost of asset replacement, if applicable	$650,000
Lost income due to time to replace asset, if applicable	$5,000
Lost income due to power loss, if applicable	$10,000
WQ issues	$0
Cost of additional treatment needed over the next year, if applicable	$0
Lost income due to lower productivity from additional treatment, if applicable $0	
Fatalities and serious injuries	$0
Total C =	**$665,000**

Vulnerability

Will the asset lose power?	1.0
Will the utility lose access to the asset?	0.5
Will the utility need to replace the asset?	0.5
Will the utility lose the source water?	0.0

Will the system demand to fight the fire cause the utility
to lose system pressure? 0.5
Will the utility require additional water treatment through
this asset? 0.0

Average V = **0.417**

Threat Likelihood

$T_{Mi} = (90,000/585,081) \times 1/1.18$

$T_{Mi} =$ **0.13036**

Risk = $\$665,000 \times 0.417 \times 0.13036$

R = \$36,149

<u>Major Wildfire</u>

Consequences

Cost of asset replacement, if applicable	$650,000
Lost income due to time to replace asset, if applicable	$5,000
Lost income due to power loss, if applicable	$12,000
WQ issues	$0
Cost of additional treatment needed over the next year, if applicable	$0
Lost income due to lower productivity from additional treatment, if applicable $0	
Fatalities and serious injuries	$0
Total C =	**$667,000**

Vulnerability

Will the asset lose power?	1.0
Will the utility lose access to the asset?	0.5
Will the utility need to replace the asset?	0.5
Will the utility lose the source water?	0.0
Will the system demand to fight the fire cause the utility to lose system pressure?	0.5
Will the utility require additional water treatment through this asset?	0.0
Average V =	**0.417**

Threat Likelihood

$T_{Ma} = (90{,}000/585{,}081) \times 1/36$

T_{Ma} **= 0.00427**

Risk = $\$667{,}000 \times 0.417 \times 0.00427$

R = \$1,188

Total Risk = \$36,149 + \$1,188 = \$37,337

This page intentionally blank.

This page intentionally blank.

This page intentionally blank.

This page intentionally blank.

This page intentionally blank.